Extractive Bioconversions

Bioprocess Technology

Series Editor

W. Courtney McGregor

Xoma Corporation
Berkeley, California

Extractive Bioconversions

edited by

Bo Mattiasson • Olle Holst

Chemical Center
University of Lund
Lund, Sweden

MARCEL DEKKER, INC.　　　New York · Basel · Hong Kong

Library of Congress Cataloging-in-Publication Data

Extractive bioconversions / edited by Bo Mattiasson, Olle Holst.
    p. cm. -- (Bioprocess technology ; v. 11)
   Includes bibliographical references and index.
   ISBN 0--8247-8272--0 (acid--free paper)
   1. Extraction (Chemistry) 2. Biotechnology---Technique
I. Mattiasson, Bo II. Holst, Olle. III. Series.
TP248.25.E88D97   1991
660'.63-- --dc20   90--49150
                                     CIP

This book is printed on acid-free paper.

MARCEL DEKKER, INC.
270 Madison Avenue, New York, New York 10016

Current printing (last digit):
10 9 8 7 6 5 4 3 2 1

PRINTED IN THE UNITED STATES OF AMERICA

Series Introduction

Bioprocess technology encompasses all of the basic and applied sciences as well as the engineering required to fully exploit living systems and bring their products to the marketplace. The technology that develops is eventually expressed in various methodologies and types of equipment and instruments built up along a bioprocess stream. Typically in commercial production, the stream begins at the bioreactor, which can be a classical fermentor, a cell culture perfusion system, or an enzyme bioreactor. Then comes separation of the product from the living systems and/or their components followed by an appropriate number of purification steps. The stream ends with bioproduct finishing, formulation, and packaging. A given bioprocess stream may have some tributaries or outlets and may be overlaid with a variety of monitoring devices and control systems. As with any stream, it will both shape and be shaped with time. Documenting the evolutionary shaping of bioprocess technology is the purpose of this series.

Now that several products from recombinant DNA and cell fusion techniques are on the market, the new era of bioprocess technology is well established and validated. Books of this series represent developments in various segments of bioprocessing that have paralleled progress in the life sciences. For obvious proprietary reasons, some developments in industry, although validated, may be published only later, if at all. Therefore, our continuing series will follow the growth of this field as it is available from both academia and industry.

W. Courtney McGregor

Preface

Biotechnological processes were long regarded as a sequence of different unit operations. The goal was then to optimize each individual step as far as possible. However, since biology in itself contains many levels of regulation, this was not the optimal way to carry out processes in all cases.

By integration, some of the unit operations are carried out simultaneously or intermittently in a coupled way. Since product inhibition and product instability are two major limitations of biotechnological processes, it was quite natural to look for synergies by integrating bioconversion and downstream processing.

This volume highlights these new developments. In doing so, we deliberately excluded the area of immobilized biocatalysts, although processes based on such technology may fall within the area of extractive bioconversions. Today, there are extensive reviews and books covering this area, although in most cases the emphasis is not on integration.

Many separation technologies are described, some in separate chapters, some as examples under a wider topic. New, interesting separation techniques that are just about to be accepted in biotechnological work have been included in our effort not only to describe the present status, but also to illustrate future paths. Examples of this are extractive bioconversions in supercritical solutions and the use of ion-pair extraction for separation.

Concerning applications, most experience is available from work with microorganisms. Therefore, this volume focuses on this topic. However, the principles of integration may also be applied to other fields of biotechnology (a separate volume on hybridoma cell cultivation has been published in this series).

Many combinations of processes and separation steps may be constructed. In this volume we draw attention to the many possibilities that extractive bio-conversions offer to biochemical engineering, by inviting leading scientists from different disciplines to describe their techniques, results obtained, and forecasts for future development.

The chapters therefore differ widely: some are theoretical, others more descriptive. It is our hope that this volume will stimulate more thinking and more work along the lines of process integration.

Bo Mattiasson
Olle Holst

Contents

Contributors

Harvey W. Blanch Department of Chemical Engineering, University of California at Berkeley, Berkeley, California

Joaquim M. S. Cabral Department of Chemical and Biochemical Engineering, Biochemical Engineering Laboratory, Instituto Superior Técnico, Universidade Técnica de Lisboa, Lisbon, Portugal

Magnus Danielsson Chematur AB, Karlskoga, Sweden

Lars Ehnström Chematur AB, Karlskoga, Sweden

Hansruedi Felix AGRO Research, Sandoz Agro Ltd., Basel, Switzerland

Jonas Frisenfelt Chematur AB, Karlskoga, Sweden

Wilhelm Gudernatsch Fraunhofer-Institut für Grenzflächen- und Bioverfahrenstechnik, Stuttgart, Federal Republic of Germany

Olle Holst Department of Biotechnology, Chemical Center, University of Lund, Lund, Sweden

Rajni Kaul Department of Biotechnology, Chemical Center, University of Lund, Lund, Sweden

Jorge L. López Sepracor Inc., Marlborough, Massachusetts

Stephen L. Matson Sepracor Inc., Marlborough, Massachusetts

Masatoshi Matsumura Institute of Applied Biochemistry, University of Tsukuba, Tsukuba, Ibaraki, Japan

Bo Mattiasson Department of Biotechnology, Chemical Center, University of Lund, Lund, Sweden

Douglas A. Miller Department of Chemical Engineering, University of California at Berkeley, Berkeley, California

Bengt-Arne Persson AB Hässle, Mölndal, Sweden

John M. Prausnitz Department of Chemical Engineering, University of California at Berkeley, Berkeley, California

John A. Quinn Department of Chemical Engineering, University of Pennsylvania, Philadelphia, Pennsylvania

Theodore W. Randolph Department of Chemical Engineering, University of California at Berkeley, Berkeley, California

Steven R. Roffler Department of Chemical Engineering, University of California at Berkeley, Berkeley, California

Göran Schill Department of Analytical Pharmaceutical Chemistry, Biomedical Center, University of Uppsala, Uppsala, Sweden

Thomas J. Stanley General Electric Company, Schenectady, New York

Heinrich Strathmann Institut für Chemische Verfahrenstechnik, Universität Stuttgart, Stuttgart, Federal Republic of Germany

Jon Sundquist Department of Chemical Engineering, University of California at Berkeley, Berkeley, California

Charles R. Wilke Department of Chemical Engineering, University of California at Berkeley, Berkeley, California

Objectives for Extractive Bioconversion

Bo Mattiasson and Olle Holst

*Chemical Center, University of Lund
Lund, Sweden*

I. DEFINITION

The term "extractive bioconversion" is synonymous with in situ product recovery. Both terms deal with the concept of increasing the productivity or performance of biotechnological processes by the continuous removal of a product from the site of its production.

II. INTRODUCTION

Biotechnological processes employ a combination of biology and technology to construct an efficient process. The biological part of such a system takes advantage of the diversity of conditions under which specialized microorganisms operate successfully. The technological part, of course, involves optimizing the process by creating favorable operating conditions and eliminating, as far as possible, some of the constraints exerted by the biology of the organism of choice.

Several levels of metabolic control have developed throughout evolution. Reversibility of enzymes, feedback allosteric control, gene regulation, permeability of cell membranes, and other features all help to control the metabolic flows in proper pathways under controlled conditions. One characteristic of these control levels is that they are sensitive to enhanced concentrations of certain metabolites. In other words, when the concentration of a metabolite increases, a whole series of control mechanisms are set into operation.

III. NEED FOR IMPROVEMENTS IN PROCESS DESIGN

A typical biotechnological process is characterized by rather low productivity in comparison to what is normal for chemical synthetic reactions. Furthermore, the product stream is dilute, which leads to high costs in the subsequent isolation and purification of the product.

From the technological point of view a maximal product concentration is desirable, whereas biology may have many mechanisms that counteract such increases. Bioprocess technology aimed at optimizing productivity therefore addresses the task of increasing productivity by means of technological solutions that enable circumvention of the biological limitations.

There are several ways to improve the overall behavior of a biotechnological process: the cells can be modified so that the product is excreted prior to reaching inhibiting concentrations, selective mutation of the cell can be employed to eliminate points of metabolic pathways sensitive to feedback inhibition, or the concentration of an enzyme of critical importance to the whole metabolic pathway can be amplified. This last point is as yet pure speculation; very little is known about what happens to the cellular physiology as a result of such changes.

Product elimination can be achieved by reacting the product in a subsequent step to transfer it to a form that does not influence the equilibrium (see Chapter 9), or it can be removed from the site of catalysis by permeabilization (see Chapter 11) or by selective elimination by membrane filtration processes (Chapters 2, 4, and 5), by extraction (Chapters 3, 6, and 7), by evaporization (Chapters 4 and 10), or by adsorption (Chapter 8).

Another approach to improving the productivity of a process is to increase the catalyst density in the reactor. This can be achieved by immobilization (1-5) and by catalyst recycling (Chapters 2 and 13).

In high-density catalytic processes the demand for efficient process control is high. Development is underway, but no real working system has yet been presented.

The present volume deals with various aspects of integration of bioconversions and the first steps in downstream processing, a technology that may lead to the development of a totally new breed of bioreactors.

In this chapter we discuss the different mechanisms by which the level of metabolites is controlled in nature and how they influence the biotechnological process; in addition, various technical solutions for the circumvention of these biological constraints are discussed. The rest of the volume explores various techniques as applied to the continuous removal of products during the process, a procedure called extractive bioconversion or in situ product recovery.

IV. INHIBITION PATTERNS

The metabolic control of metabolite levels may be exerted on different levels. The best known is reversibility in the enzyme-catalyzed process leading to product inhibition. Here, direct control occurs when an enrichment of products takes place. This type of control is especially important for reactions running close to equilibrium, for example isomerase-catalyzed reactions (6) and many enzymatic steps catalyzed by NAD (P)-dependent dehydrogenases (7). In feedback control, however, the product from one enzyme-catalyzed step interferes with another enzyme activity several steps earlier in the metabolic pathway. Well-known examples of this type of metabolic control are found in the synthesis of aromatic amino acids (8). Yet another type of control of catalytic activity is substrate inhibition. Here, the enzyme catalysis is affected by allosteric control as a result of more than one type of binding site for the substrate molecule on the enzyme. These types of metabolic control are schematically illustrated in Fig. 1.

The systems are exemplified by well-defined enzymes; this is because such systems are the best known. Of course one can foresee similar behavior for transport proteins carrying out the transport of substrate and product molecules across membrane barriers, for example.

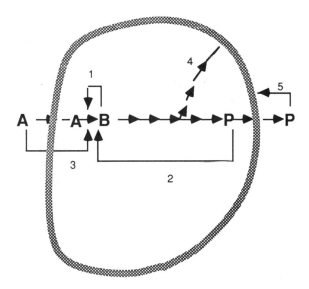

Figure 1 Schematic presentation of the different types of inhibition taking place in a bioconversion process: (1) product inhibition; (2) feedback inhibition; (3) substrate inhibition; (4 and 5) unspecific inhibition by by-products.

Besides effects on the molecular level, there are many examples in which enhanced levels of metabolites can drastically change the performance of a bioconversion process. When producing organic acids or alcohols severe product inhibition has been reported. These effects may to a large extent be ascribed to effects on membrane permeability properties caused by these products (9-11). Thus, the metabolite causes a more general influence on the cell and does not specifically affect the enzyme(s) essential for the biocatalysis leading to the formation of the product. Weak acids may, upon interfering with the membranes, cause a depletion of the pH gradient across the membrane. Since this gradient is believed to be responsible for the phosphorylation of ADP, such effects may cause severe disturbances in the energy supply to the cells and thus switch the metabolism in new directions. From model studies, it is known that when the phosphorylation of ADP in *Saccharomyces cerevisiae* is inhibited, an increased rate of glycolysis is observed (12-14).

It may sometimes be difficult to clearly differentiate between the different types of negative effects on the biocatalyst. The more simple the biocatalyst, the easier this differentiation becomes. When dealing with pure enzymes, free or immobilized, a fairly good theoretical background exists for both molecular properties per se and microenvironmental effects (15). The theoretical background for the knowledge of events on a cellular level is far less developed. Microbial physiology is an underdeveloped discipline, and much more attention must be directed to this issue before full potential can be reached by improvements in fermentation technology and process control. For immobilized cells, the work of identifying the rules of the game has just begun (16).

Another area that is poorly understood is that of plant cell biology. In many cases the vacuole has been regarded as a sink for many substances produced by the cell, products that have been difficult to excrete. Sometimes these stored molecules are attractive from the biotechnological point of view. It would be desirable to change the transport routes in the cell and cause excretion of the molecules to the surrounding medium instead. This problem has not been generally solved. However, in the area of cell permeabilization (see Chapter 11), it has now been shown that it is possible to change the permeability properties to such an extent that the vacuole can be emptied of its contents of storage molecules without ruining the cell metabolism. This has been interpreted in terms of a temporary permeabilization that ceases after treatment. Membrane properties are restored, and metabolism can continue as before. By the use of adsorbents in the medium some changes in the transport directions in the cell have been reported (17).

Besides the fact that the production rate may be reduced by a buildup of product concentrations, there are other reasons for running extractive fermentations. Some of these are discussed in the following sections.

V. PRODUCT STABILITY

When product is excreted into the medium, the often hostile conditions may efficiently reduce the yield of active substance if the product is not removed immediately. The bubbling of air or oxygen in aerobic reactions, the shear forces generated by vigorous stirring, and the presence of hydrolytic enzymes are three strong reasons for efficiently and rapidly removing the product from the medium. In this volume the elimination of sensitive antibiotics (cycloheximide) by the use of solid sorbents directly in the fermentor is discussed (18).

The production of extracellular proteins is not problematic when working with conventional strains: the proteases from one organism do not attack other exoenzymes from the same organism. However, when introducing foreign gene material into organisms and expressing new proteins, the old proteases become a severe threat to the new proteins. Thus far the solution has been to eliminate the genes for proteases as efficiently as possible without endangering the welfare of the cell.

Another appealing alternative is to trap the target molecule by a specific sorbent as soon as it is excreted or to render the proteases ineffective by applying protease inhibitors. These inhibitors may be added either free in solution or bound to solid sorbents. This latter approach is more general when a broad-spectrum protease inhibitor preparation is used. This has not been reported in practice yet to any large extent. However, there may be some activity along these lines in the area of mammalian cell culturing.

Yet another reason to eliminate a product from a culturing system is when the product is hazardous to the cell itself. Examples that have been frequently cited in literature are those of the genus *Clostridium*. These cells produce auto-lysins, that is, proteases that cause autolysis of the cells (19,20). To avoid this, the now-classic experiment by Hedén and Puziss, who cultivated *Clostridium tetanii* and developed a much more efficient cultivation system by continuously extracting the toxin by means of an aqueous two-phase system (21).

When the product is an intermediary metabolite in a metabolic pathway, it is recommended that it be removed before large concentrations build up. In such situations one often operates with mutants lacking the subsequent enzyme in the sequence, thereby making it possible to enrich the metabolite of interest. The mutant may be fully deficient in the subsequent enzyme activity, or the enzyme may be severely disturbed. In the latter case very high substrate concentrations may initiate catalytic activity as well as starting feedback inhibition processes. Therefore, it is highly desirable to remove the product from the site of production. This is performed by reducing the content in the broth, thereby making it more favorable to excrete the product from the cell. Another attractive alternative is to facilitate elimination from the cell interior by means of permeabilization (see Chapter 11).

When developing extractive bioconversion processes it is of the utmost importance to select extraction media that do not interfere with the cells, or do so in a controlled and desirable way (see Chapters 6, 7, and 9). Here, one must differentiate between unspecific membrane permeabilization and toxicity effects on the cell. Organic solvents have been regarded as problematic in this respect, but more recent research clearly demonstrates the possibility of finding suitable organic extractants (22,23). More recent extraction systems, such as supercritical carbon dioxide (Chapter 6) and aqueous two-phase systems (Chapter 7), are regarded as more gentle to the cells.

VI. TECHNICAL CONSTRAINTS

That biotechnological processes by nature are slow and self-regulating and result in dilute aqueous product streams has caused a technical development toward more highly producing systems involving increased cell densities, better process control, and controlled substrate delivery. Controlled product removal, which is one of the aims of extractive bioconversion, adds yet another dimension to this process technology. These new conditions of course amplify many of the constraints mentioned earlier in this chapter and raise demands for efficient ways to handle them. This volume presents a wide variety of techniques applied in integrating bioconversion with downstream processing. We have deliberately excluded enzymes and cells immobilized in solid supports. This is done despite that product is continuously removed by the product stream. Many volumes have been published dealing with this matter, and we therefore refer to those (1–5). However, some experience from immobilized cells is discussed in relation to the other technologies treated in this volume.

When operating with high cell densities (in cell recycling, for example) substrate may be fed continuously and product removed continuously. Such an arrangement may cause problems; for example, some products may not be removed along with the product stream. Then there is an enrichment of by-products and eventually these affect the process, for example when glucose is fed to a *Saccharomyces cerevisiae* culture for production of ethanol. If the ethanol is removed by distillation, any glycerol produced with the ethanol remains in the medium and is recirculated back to the fermenter (24). Increasing concentrations of glycerol change the osmotic environment of the cells. A way around this problem is to operate with a bleed stream by which excess cells and some by-products may be removed. When operating with more complex media, such as molasses, nonbiodegradable compounds may be present and recirculated in the fermentation broth. A bleed stream is also the solution to problems of enrichment in this case.

Another situation in which it is advantageous to be able to remove the product stream continuously is when cells must be cultured at dilution rates that under conventional cultivation conditions would cause washout of the fermenter. For example, this may be necessary when growing cells on toxic substrates that can only be administered in very low concentrations to avoid intoxication of the cells. When operating with fed-batch cultivation one can, of course, also envision a feed with concentrated substrate, but under such conditions it is highly probable that hot spots will occur in the fermentor and cause cell damage. Feeding a dilute substrate concentration would then be more favorable. Here, however, water must be removed to increase the cell density above what was otherwise attainable with the substrate concentration used. An example of this type of cultivation is given in the chapter on cell culture under membrane recirculation (Chapter 2), in which *Pseudomonas cepacia* was grown on salicylate in a membrane-filtered culture broth (25).

As stated earlier, knowledge of microbial physiology is of utmost importance to the full exploitation of the potential that new technological solutions offer. By increasing the cell density in immobilized preparations or in cell recycling systems, for example, the cells are subject to conditions of stress. This may be because of a stronger competition for the oxygen available, and it may lead to situations in which high concentrations of polymer material bind enough water that the water activity is markedly changed. This kind of stress on cells has been observed and reported (26-28), and among other problems leads to changes in the product pattern of an organism. So far, very few attempts to explain the underlying relations have been published (16), but it is obvious that here is an additional way of influencing cell metabolism and thus the product pattern of the process.

When particulate or macromolecular or generally less soluble substrates are to be converted, extractive bioconversions may be an attractive alternative (29). Product inhibition or reversible reactions make it attractive to efficiently remove the product, and this is also true when dealing with poorly soluble substances.

In summary, there may be reasons for performing extractive bioconversion because of enzyme properties or cellular regulation or because of effects of the technological solutions taken.

ACKNOWLEDGMENTS

Part of this work was supported by the National Swedish Board for Technical Development.

REFERENCES

1. Salmona, M., Saronio, C., and Garattini, S., eds., *Insolubilized Enzymes*. Raven Press, New York (1974).

2. Chang, T. M. S., ed., *Biomedical Applications of Immobilized Enzymes and Proteins*, Vols. I and II. Plenum Press, New York (1977).
3. Mosbach, K., ed., *Methods Enzymol.* 44 (1976).
4. Mattiasson, B., ed., *Immobilized Cells and Organelles*, Vols. I and II. CRC Press, Boca Raton, FL (1983).
5. Mosbach, K., ed., *Methods Enzymol.* 137 (1988).
6. Rugh, S., Industrial-scale production and application of immobilized glucose isomerase, *Methods Enzymol.* 136:356–370 (1987).
7. Lee, L. G., and Whitesides, G. M., Preparation of optically active 1,2-diols and α-hydroxyketones using glycerol dehydrogenase as catalyst: Limits to enzyme-catalyzed synthesis due to non-competitive and mixed inhibition by product, *J. Org. Chem.* 51:25–36 (1986).
8. Moyed, H. S., and Umbarger, H. E., Regulation of biosynthetic pathways, *Physiol. Rev.* 42:444–466 (1962).
9. Maiorella, B., Blanch, H. W., and Wilke, C. R., By-product inhibition effects on ethanolic fermentation by *Saccharomyces cerevisiae*, *Biotechnol. Bioeng.* 25:103–121 (1983).
10. Ingram, L. O., Microbial tolerance to alcohols: Role of the cell membrane, *Trends Biotechnol.* 4:40–44 (1986).
11. Herrero, A. A., End-product inhibition in anaerobic fermentations, *Trends Biotechnol.* 1:49–53 (1983).
12. Hahn-Hägerdal, and Mattiasson, B., Azide sterilization of fermentation media. Ethanol production from glucose using immobilized *Saccharomyces cerevisiae*, *Eur. J. Appl. Microbiol. Biotechnol.* 14:140–143 (1982).
13. Pinto da Costa, S. O., and Bacila, M., Induction of respiratory deficient nonchromosomal "petites" of *Saccharomyces cerevisiae* by sodium dodecyl sulphate, *J. Bacteriol.* 115:461–463 (1973).
14. Lehninger, A., *Biochemistry*, 2nd ed. Worth, New York (1975).
15. Goldstein, L., Kinetic behaviour of immobilized enzyme systems, *Methods Enzymol.* 44:397–443 (1976).
16. Mattiasson, B., and Hahn-Hägerdal, B., Microenvironmental effects on metabolic behaviour of immobilized cells. A hypothesis, *Eur. J. Appl. Microbiol. Biotechnol.* 16:52–55 (1982).
17. Payne, G. F., Payne, N. N., Schuler, M. L., and Asada, M., In situ adsorption for enhanced alkaloid production by *Catharanthus roseus*, *Biotechnol. Lett.* 10:187–192 (1988).
18. Wang, H. Y., Integrating biochemical separation and purification steps in fermentation processes. *Ann. N. Y. Acad. Sci.* 413:313–321 (1983).
19. Soucaille, P., Joliff, G., Izard, A., and Goma, G., Butanol tolerance and autobacteriocin production by *Clostridium acetobutylicum*, *Curr. Microbiol.* 14:295–299 (1987).
20. Barber, J. M., Robb, F. T., Webster, J. R., and Woods, D. R. Bacteriocin production by *Clostridium acetobutylicum* in an industrial fermentation process, *Appl. Environ. Microbiol.* 37:433–437 (1979).
21. Puziss, M., and Hedén, C.-G., Toxin production by *Clostridium tetani* in biphasic liquid cultures, *Biotechnol Bioeng.* 7:355–366 (1965).

22. Brink, L. E. S., and Tramper, J., Optimization of organic solvent in multiphase biocatalysis, *Biotechnol. Bioeng.* 27:1258–1265 (1985).
23. Laane, C., Boeren, S., and Vos, K., On optimizing organic solvents in multiliquid phase biocatalysis. *Trends Biotechnol.* 3:251–252 (1985).
24. Kühn, I., Alcoholic fermentation in an aqueous two-phase system, *Biotechnol. Bioeng.* 22:2393–2398 (1980).
25. Berg, A-C., Holst, O., and Mattiasson, B., Continuous culture with complete cell recycle; cultivation of *Pseudomonas cepacia* ATCC 29351 on salicylate for production of salicylate hydroxylase, *Appl. Microbiol. Biotechnol.* 30: 1–4 (1989).
26. Kopp, B., and Rehm, H. J., Alkaloid production by immobilized mycelia of *Claviceps purpurea, Eur. J. Appl. Microbiol. Biotechnol.* 18:257–263 (1983).
27. Doran, P. M., and Bailey, J. E., Effects of immobilization on growth, fermentation properties and macromolecular composition of *Saccharomyces cerevisiae* attached to gelatin, *Biotechnol. Bioeng.* 28:73–87 (1986).
28. Doran, P. M., and Bailey, J.E., Effects of immobilization on the nature of glycolytic oscillations in yeast, *Biotechnol. Bioeng.* 29:892–897 (1987).
29. Larsson, M., Arasaratnam, V., and Mattiasson, B., Integration of bioconversion and down-stream processing. Starch hydrolysis in an aqueous two-phase system, *Biotechnol. Bioeng.* 33:758–766 (1989).

2

Cultivation Using Membrane Filtration and Cell Recycling

Olle Holst and Bo Mattiasson

Chemical Center, University of Lund
Lund, Sweden

I. INTRODUCTION

Membrane technology has been applied in many different areas of biotechnology (Michaels, 1980; McGregor, 1986; Strathmann, 1985; Gekas, 1986). The most obvious application in which most large-scale processes are also found is in downstream processing. The integration of membrane separation and bioconversion is very attractive, and several examples of extractive bioconversions are known at present (see also Chapters 3–5).

This chapter focuses on the use of membranes for cell recycling in fermentation processes. The incentive for this integration may be a desire to harvest products formed in the cell suspension while the cells continue production, or integration may be a way to increase cell density in a fermentor while the low-molecular weight products are removed. These two reasons may at first glance appear to be very similar, the only difference being the product recovered. However, there may be different strategies behind these two alternatives. With low-molecular-weight products, the only important objectives are productivity and yield, that is, the amount of product formed per unit time and volume and the amount of product formed per amount of substrate consumed, respectively. In the case of cell mass production, a quality aspect may also be of importance for the eventual application.

The practical separation of low-molecular-weight compounds in the broth from the cells may be carried out by using either an external loop (Fig. 1a)

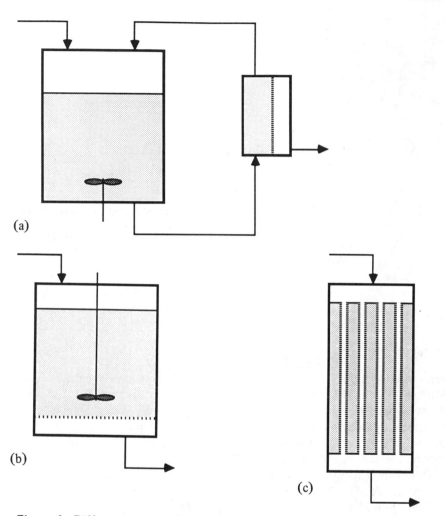

Figure 1 Different modes of operating membrane reactors: (a) external membrane loop, (b) internal membrane, (c) immobilization in hollow-fiber unit.

equipped with a membrane unit or a filter device mounted inside (Fig. 1b) the fermentor (Dosta'lek and Häggström, 1982).

An extreme variation of the fermentor with an internal dialysis unit is the use of a hollow-fiber reactor for culturing cells on the shell side of the membrane (Fig. 1c). Substrate is then fed to the cells and products removed via the lumen of the fiber (Vick Roy et al., 1983a). A limiting factor may then be mass transfer over the membrane unit.

When operating with anisotropic membranes, it has been demonstrated that the cells cultured grow as close to the substrate source as possible. Therefore a local high concentration is generated in the grooves of the individual anisotropic membrane fibers. When the cells continue to divide, the pressure rises and eventually the membrane bursts, releasing cells into the substrate feed (Mattiasson et al., 1981; Inloes et al., 1983). This kind of reactor is best regarded as a variation of immobilized cells (Mattiasson and Larsson, 1985) and is not covered further in this chapter.

The two methods based on external or internal membranes in conjunction with the fermentation unit both have their advantages and drawbacks. The external loop can easily be changed if severe clogging occurs. Furthermore, it is easy to increase the filtration area when operating outside the fermentor. When operating with high cell densities using an external loop there is an obvious risk of the occurrence of oxygen limitation. One way to reduce these risks is to increase the speed of circulation through the membrane unit, thereby reducing the delay in the membrane unit.

Adsorption or attachment of cells to the membrane surface is a problem regardless of how the membrane is arranged. Back-flushing may restore the flux through the membrane, at least partially.

II. CONTINUOUS CULTURE WITH COMPLETE CELL RECYCLING FOR PRODUCTION OF CELL MASS

Complete cell recycling for improving the cell density may be especially attractive when dealing with product- as well as substrate-inhibited cell cultures. At present there is very little basic knowledge about such cultivation systems, and therefore one must take into consideration whether the cultivation conditions influence the cells in a negative way. Several potential technical and biological problems may occur:

Problems in supplying sufficient nutrients in a well-controlled way to a dense cell suspension

Changes in yield coefficient due to changes in growth rate and to a changed proportion of the substrate being used for maintenance

Adhesion of the cultured cells to the membrane used

Mechanical disturbance of the cells by pumping and severe shear stress

Changes in product formation pattern due to changes in growth rate

Oxygen limitations in the external membrane loop during recycling broth with high cell densities

Oxygen limitations in the fermentor at high cell densities

With such a long list of potential complications, it soon becomes obvious that either not all complications will occur or the advantages realized with the technology are great: otherwise nobody would try to apply membrane recirculation techniques.

A. Problems in the Proper Supply of Feed to a Dense Cell Suspension

When cultivating *Streptococcus lactis* for production of the intracellular enzyme superoxide dismutase (E.C. 1.15.1.1), cell recycling with a concomitant removal of lactate was a necessity (Holst et al., 1985). Severe inhibition occurred at levels as low as 8 g lactate per L, and only approximately 2 g cell mass (dry weight) was obtained per L fermentation broth. If a membrane filtration step was added after the fed batch, lactate was kept at a constant level even when fresh medium was continuously fed to the system. When feeding at a constant rate, the cell mass rose to approximately 13 g/L in 60 h (Fig. 2). It should be stressed that if the feed rate is sufficient in the initial phase of the cell recirculation it soon becomes too low when cell mass increases. This is observed as a successively slower increase in cell density despite a constant feed rate. It is possible that a successively increasing percentage of the substrate is either not metabolized or is used for maintenance metabolism of the cells.

By manually changing the feed rate a cell density of 18 g/L was achieved in 20 h. By emptying the fermentor of all but approximately 10% of the broth, a new feed batch and subsequent membrane filtration cultivation could be carried out. This example clearly illustrates the need for improved process control when operating with increased cell densities.

B. Yield Coefficient

The technique of complete cell recycling has been suggested as a method for determining the maintenance coefficient (Beyeler et al., 1984). When cultivating *Zymomonas mobilis* in a system with complete cell recycling, Beyeler and co-workers were able to estimate the maintenance coefficient to approximately 1.6 g/g·h. This figure should be compared to data from continuous culture in which

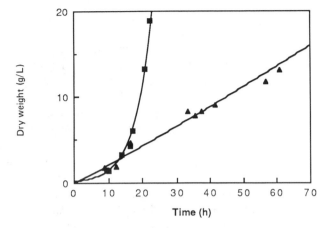

Figure 2 Cultivation of *Streptococcus lactis* in continuous culture with complete cell recycling: (▲) constant dilution rate, and (■) increased dilution rate. (Data from Holst et al., 1985.)

the maintenance coefficient was estimated as approximately 1.5 g/g·h. The results suggest that the technique used did not affect the microorganisms negatively.

Even if the cells per se are not negatively influenced by the cultivation as indicated by maintenance studies, it is obvious that the high cell densities result in one of two conditions: either a less efficient supply of nutrients to the cells, thereby causing more of the substrate to be used for maintenance, or the high cell density causing some kind of stress situation, leading to a greater need for maintenance energy, which is observed as a lower yield in the process.

C. Adhesion of the Cultured Cells

When culturing cells in continuous processes under high dilution rates there is a tendency for the cells to become stickier and they have a stronger tendency to attach to solid surfaces (Molin et al., 1982). The same kind of behavior is observed under very limited substrate conditions (Keffort et al., 1982).

When culturing an organism on a toxic substrate, there are several similarities to the situations just described. The cells are cultured under dilution rates that otherwise would cause a washout of the fermentor, and the substrate level is deliberately kept low to avoid locally high concentrations of substrate that could cause unwanted effects on the cells.

In an experiment in culturing *P. cepacia* in a membrane recirculating system, the carbon source used, salicylate, was toxic above a concentration of 2 g/L. Such a low concentration gave a yield of only approximately 0.6 g of cell mass per L. When starting membrane recirculation after the salicylate in the initiating batch cultivation was depleted, the cell concentration in the medium stayed very low for quite some time (Fig. 3). This was later shown to be due to adhesion to the membrane surfaces in the filtration unit. Not until the membrane unit was saturated did the cell density in the fermentor system start to raise (Berg et al., 1989).

P. cepacia was cultivated to produce the enzyme salicylate oxygenase (EC 1.14.13.1). When studying the yield of enzyme as well as the specific activity it was clear that, even if the data seem a bit irregular, there are no drastic changes in specific activity under the cultivation conditions (Fig. 3). This means that one can expect a higher total productivity of the enzyme with increasing cell mass production.

This study may be a good model for the cultivation of many of the organisms that today attract much attention because of their ability to degrade toxic compounds and that are expected to have a bright future in environmental biotechnology.

D. Are the Cells Too Brittle for This Treatment?

When pumping the fermentation broth through the membrane loop, the cells are exposed to rather harsh conditions. Pumping per se may be harmful to the cells, as well as the shear forces obtained when the broth is forced through the filter unit at high speed. To investigate the effects of this treatment several authors have reported studies in which the leakage of intracellular enzymes was monitored. Presumably, if the cells are hurt, there is leakage of intracellular enzymes. When simply measuring the total content of protein in the fermentation broth no effects were observed. The results were the same when analyses were made for enzyme specific to the intracellular volume (Holst et al., 1985). Thus, at least in the cases studied, the mechanical stress was not strong enough to cause severe cell damage. One should remember in this context that the cells studies are all supported by strong cell walls, and the situation may be rather different with less protected cells, such as mammalian cells.

E. Changes in the Product Pattern

When cells are exposed to stress one can observe changes in product pattern (Chapter 7) (Mattiasson and Hahn-Hägerdal, 1982). It was therefore an

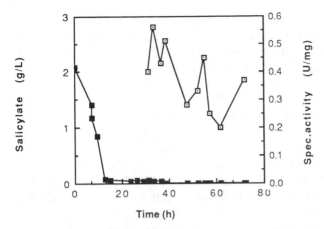

Figure 3 Cultivation of *Pseudomonas cepacia* in continuous culture with complete cell recycling: (■) concentration of salicylate, and (▣) specific activity of salicylate hydroxylase. (Data from Berg et al., 1989.)

interesting task to investigate whether there were any changes in product pattern when cultivating microbial cells under complete cell recycling in a membrane unit. In all the publications dealing with the production of low-molecular-weight products, no such effects are reported. However, Hjörleifsdottir et al. (1990) found changes in the product pattern when cultivating *Lactobacillus casei* in the sense that the amount of D-lactate formed increased in proportion to the total amount of lactate produced. When cultivation the cells on 50 g glucose per L in a batch cultivation, 3% D-lactate was formed. However, when operating with cell recycling the amount increased from 3.8 to 6% during a period of 129 h. In the meantime the cell mass increased from 6 to 80 g dw per L. Similar results were obtained when cultivating on 25 g/L of glucose. It deserves mentioning that in a control experiment when a batch culture was starved a substantial increase in the percentage of D-lactate was observed. The membrane recirculation culture thus behaves similarly to the starved batch culture. The operation at high cell densities puts some stress on the cells, and it may therefore be expected that some kind of stress effect is expressed in the metabolic behavior.

F. Change in Metabolic Status

Metabolic stress and changes in metabolic behavior must be a result of changes within the cell. To investigate whether it was possible to observe any such effects, studies on the protein composition were performed (Hjörleifsdottir et al., 1990). By running two-dimensional gel electrophoresis it was possible to obtain a kind of "fingerprint" of the cells under the different cultivation conditions. From these studies it was obvious that several new proteins appeared, that some proteins disappeared, and that the amounts of several other proteins changed drastically. It is thus likely that the intracellular situation is quite drastically changed when operating with microorganisms under stress.

Similar results were obtained by Hoffmann and coworkers (1987), who studied cell mass retention of *Saccharomyces cerevisiae* using laser flow cytometry. When increasing the dilution rate above 0.28 h^{-1}, they found that yeast cells changed from oxidative to fermentive metabolism. Furthermore, when increasing the dilution rate even further to D = 0.39 h^{-1} they found interesting changes in protein content, RNA content, DNA content, and cell size. The size and DNA content became smaller with continued recycling. An possible explanation is that a decreasing fraction of the cells were growing and budding. From this point of view it is surprising that no more such observations have yet been reported.

When culturing *Lactococcus lactis* under similar conditions, it was observed that the cells changed from homofermentative to heterofermentative metabolism (Hjörleifsdottir et al., in press).

G. Oxygen Limitations in External Loop or Fermentor

The supply of oxygen to aerobic cells is one of the classic problems in fermentation technology. When increasing the cell density these problems are accentuated. There is as yet no general rule for how to improve oxygen supply, even if there are many alternative techniques (Enfors and Mattiasson, 1983; Holst, 1984; Adlercreutz, 1985). The supply in the fermentor is a slightly easier problem since this is a function of aeration. In the external loop oxygen deficiency very rapidly appears. There are some possibilities for improving oxygen supply by working with sandwich-layer membranes in which the cell-containing broth is passed close to a layer carrying a high content of oxygen. In this manner oxygenation may take place virtually as in an membrane oxygenator.

The most obvious applications of the membrane recirculation technology are for anaerobic cells or facultative anaerobes. Such constraints as insufficient gas transfer may not be a problem in these systems.

The general strategy for dealing with the problems of insufficient oxygen transfer is first to try to supply as much oxygen as possible; second, to reduce the length of time when the cells are exposed to unfavorable conditions; and finally not to operate with very high cell densities.

III. PRODUCTION OF LOW-MOLECULAR-WEIGHT PRODUCTS

The same constraints as already discussed are also valid in the production of low-molecular-weight products. However, thus far only processes that operate under anaerobic conditions have been employed. This means that oxygen supply has not been a problem. The technology is otherwise in many respects identical.

In Tables 1 and 2 are listed some of the examples found in the literature of the production of low-molecular-weight metabolites. It is clear that the productivity may be very high in some reports, but if the goal is the development of processes in which the substrate is utilized as efficiently as possible, the amount of substrate remaining in the spent broth must be taken into consideration.

Furthermore, in these processes the slow removal of cell mass by a bleed stream is advantageous. Using such an arrangement both the old cell mass and unwanted by-products or impurities in the feed may be removed. A bleed stream is necessary when product formation is coupled to cell growth.

A number of studies of cell mass retention have been carried out in various systems. Among the products are ethanol, acetone, butanol (Pierrot et al., 1986; Afschar et al., 1985), lactic acid (Vick Roy et al., 1983b), propionic acid (Boyaval and Corre, 1987), citric acid (Enziminger and Asenjo, 1986), L-sorbose (Bull and Young, 1981), and 2-ketogluconic acid (Bull and Young, 1981). Further examples and some of the characteristics of the processes studied are summarized in Table 3.

IV. SELECTION OF MEMBRANE

There is not yet sufficient information to permit any clear judgment concerning the selection of membrane type. Ultrafiltration membranes as well as microfilters have been used. Microfilters have a higher initial flux and allow larger molecules to pass. On the other hand, because of the open structure, cells and cell debris may partially enter the pores and cause clogging. Ultrafilter membranes, in contrast, have a more closed structure that is not clogged by cells or cell debris, but a lower flux occurs.

The composition of the membrane is important for all kinds of applications. Fouling is a major problem (Defrise and Gekas, 1988). Since fermentation broth containing cells and cell debris together with proteins and various smaller

Table 1 Production of Ethanol in Membrane Recycling Fermentations

Microorganism	Max dw (g/L)	Conversion (%)	Productivity (g/L·h)	Reference
Saccharomyces bayanus	133	100	80	Mota et al, 1987
S. cerevisiae	36	85	11	Damiano et al., 1985
	210	100	85	Lee and Chang, 1987
	86	87	44	Hoffmann et al., 1987
	85	83	27	Nishizawa et al, 1983
	300	93	150	Lafforgue et al., 1987
	120	70	130	Cheryan and Mehaia, 1984
Kluyveromyces fragilis	90	60	240	Cheryan and Mehaia, 1983
Zymomonas mobilis	45	100	140	Lee et al., 1980

Table 2 Production of Lactate in Membrane Recycling Fermentations

Microorganism	Max dw (g/L)	Conversion (%)	Productivity (g/lh)	Reference
Lactobacillus delbreuckii	54	100	76	Vick Roy et al., 1983b
	140	93	160	Ohleyer et al., 1985
L. bulgaricus	~60	99	85	Mehaia and Cheryan, 1986
L. helveticus	64	<100	22	Boyaval et al., 1987

Table 3 Examples of Fermentations Performed in Membrane Recycling Fermentations

Microorganism	Product	Max dw (g/L)	Productivity (g/lh)	Reference
Saccharomycopsis lipolytica	Citric acid	26	1.2	Enzminger and Asenjo, 1986
Streptococcus lactis	SOD	19	–	Holst et al., 1985
Propionibacterium acidipropionici	Propionic acid	112	2.1	Blanc and Goma, 1987a
	Propionic acid	100	14.3	Boyaval and Corre, 1987
Bifidobacterium longum	Cell mass	54	–	Taniguchi et al., 1987a
Pseudomonas cepacia	Salicylate hydroxylase	15	–	Berg et al., 1989
Clostridium acetobutylicum	Acetone-butanol	8	5.4	Afschar et al., 1985, 1986
	Acetone-butanol	20	6.5	Pierrot et al., 1986
	Acetone-butanol	125	4.5	Ferras et al., 1986
Gluconobacter suboxydans	L-sorbose	–	111	Bull and Young, 1981
Serratia marcescens	2-Ketogluconic acid	–	11	Bull and Young, 1981
Propionibacterium shermannii	Vitamin B_{12}	227	–	Hatanaka et al., 1988
Bacillus subtilis	Pectate lyase	8	–	Jauneau et al., 1988
Butyribacterium methylotrophicum	Vitamin B_{12}	33	–	Hatanaka et al., 1988
Streptococcus cremoris	Cell mass	82	4.5	Taniguchi et al., 1987b
Lactobacillus casei	Cell mass	49	2.2	Taniguchi et al., 1987b

molecules is a very complex medium, it is difficult to define the desirable chemical properties of an ideal membrane.

Very high cell densities have been reported when using ceramic membranes, and one can foresee that these membranes will be very useful in producing low-molecular-weight compounds.

V. CONCLUDING REMARKS

Membrane recycling cultivation offers an interesting new way to operate with high cell densities, to efficiently reduce inhibitory compounds, and to administer toxic substrates to growing cells. The changes observed in both intracellular protein pattern and product composition from the cells must be further evaluated before the full potential of this technology can be utilized.

ACKNOWLEDGMENT

This work was partially financed by the National Swedish Board for Technical Development.

REFERENCES

Adlercreutz, P., Oxygen supply to immobilized cells, Ph.D. Thesis, Department of Biotechnology, University of Lund (1985).

Afschar, A. S., Biebl, H., Schaller, K., and Schügerl, K., Production of acetone and butanol by *Clostridium acetobutylicum* in continuous culture with cell recycle, *Appl. Microbiol. Biotechnol.* 22:394–398 (1985).

Afschar, A. S., Schaller, K., and Schügerl, K., Continuous production of acetone and butanol with shear-activated *Clostridium acetobutylicum, Appl. Microbiol. Biotechnol.* 23:315–321 (1986).

Berg, A-C, Holst, O., and Mattiasson, B., Continuous culture with complete cell recycle: Cultivation of *Pseudomonas cepacia* ATCC 29351 on salicylate for production of salicylate hydroxylase, *Appl. Microbiol. Biotechnol.* 30:1–4 (1989).

Beyeler, W., Rogers, P. L., and Fiechter, A., A simple technique for direct determination of maintenance energy coefficient: An example with *Zymomonas mobilis, Appl. Microbiol. Biotechnol.* 19:277–280 (1984).

Blanc, P., and Goma, G., Propionic acid fermentation: Improvement of performances by coupling continuous fermentation and ultrafiltration, *Bioprocess Eng.* 2:137–139 (1987a).

Blanc, P., and Goma, G., Kinetics of inhibition in propionic acid fermentation, *Bioprocess Eng.* 2:175–179 (1987b).

Boyaval, P., and Corre, C., Continuous fermentation of sweet whey permeate for propionic acid production in a CSTR with UF recycle, *Biotechnol. Lett.* 9:801–806 (1987).

Boyaval, P., Corre, C., and Terre, S., Continuous lactic acid fermentation with concentrated product recovery by ultrafiltration and electrodialysis, *Biotechnol. Lett.* 9:207–212 (1987).

Bull, D. N., and Young, M. D., Enhanced product formation in continuous culture with microbial cell recycle, *Biotechnol. Bioeng.* 23:373–389 (1981).

Cheryan, M., and Mehaia, M. A., A high-performance membrane bioreactor for continuous fermentation of lactose to ethanol, *Biotechnol. Lett.* 5:519–524 (1983).

Cheryan, M., and Mehaia, M. A., Ethanol production in a membrane recycle bioreactor. Conversion of glucose using *Saccharomyces cerevisiae*, *Process Biochem.* December:205–208 (1984).

Damiano, D., Shin, C-S., Ju, N., and Wang, S. S., Performance, kinetics, and substrate utilization in a continuous yeast fermentation with cell recycle by ultrafiltration membranes, *Appl. Microbiol. Biotechnol.* 21:69–77 (1985).

Defrise, D., and Gekas, V., Microfiltration membranes and the problem of microbial adhesion. A literature survey, *Process Biochem.* August:105–116 (1988).

Dosta'lek, M., and Häggström, M., A filter fermenter—apparatus and control equipment, *Biotechnol. Bioeng.* 24:2077–2086 (1982).

Enfors, S. O., and Mattiasson, B., Oxygenation of processes involving immobilized cells, in: *Immobilized Cells and Organelles*, Vol. 2 (B. Mattiasson, ed.). CRC Press, Boca Raton, FL, pp. 41–60 (1983).

Enzminger, J. D., and Asenjo, J. A., Use of cell recycle in the aerobic fermentative production of citric acid by yeast, *Biotechnol. Lett.* 8:7–12 (1986).

Ferras, F., Minier, M., and Goma, G., Acetonobutylic fermentation: Improvement of performances by coupling continuous fermentation and ultrafiltration, *Biotechnol. Bioeng.* 28:523–533 (1986).

Gekas, V. C., Artificial membranes as carriers for the immobilization of biocatalysts, *Enzyme Microb. Technol.* 8:450–460 (1986).

Hatanaka, H., Wang, E., Taniguchi, M., Iijima, S., and Kobayashi, T., Production of vitamin B_{12} by a fermentor with a hollow-fiber module, *Appl. Microbiol. Biotechnol.* 27:470–473 (1988).

Hjörleifsdottir, S., Holst, O., and Mattiasson, B., Effects on product formation in *Lactococcus lactis* 65.1 in continuous culture with complete cell recycling, *Bioprocess Eng.* (in press).

Hjörleifsdottir, S., Seevaratnam, S., Holst, O., and Mattiasson, B., Effects of complete cell recycling on product formation pattern in *Lactobacillus casei* spp. *rhamnosus* in continuous culture, *Curr. Microbiol.* 20:287–292 (1990).

Hoffmann, H., Scheper, T., Schügerl, K., and Schmidt, W., Use of membranes to improve bioreactor performance, *Chem. Eng. J.* 34:B13–B19 (1987).

Holst, O., Improved oxygen supply to immobilized cells by in situ oxygen generation, Ph.D. Thesis, Department of Applied Microbiology, University of Lund, (1984).

Holst, O., Hansson, L., Berg, A-C., and Mattiasson, B., Continuous culture with complete cell recycle to obtain high cell densities in product inhibited cultures; cultivation of *Streptococcus lactis* for production of superoxide dismutase, *Appl. Microbiol. Biotechnol.* 23:10–14 (1985).

Inloes, D. S., Taylor, D. P., Cohen, S. N., Michaels, A. S., and Robertson, C. R., Ethanol production by *Saccharomyces cerevisiae* immobilized in hollow fiber reactor, *Appl. Environ. Microbiol.* 46:264–278 (1983).

Jauneau, A., Morvan, C., Fenyo, J. C., and Demarty, M., Pectate lyase production by *Bacillus subtilis* in a membrane reactor, *Appl. Microbiol. Biotechnol.* 29:234–238 (1988).

Keffort, B., Kjelleberg, S., and Marshall, K. C., Bacterial scavenging: Utilization of fatty acids localized at a solid-liquid interface, *Arch. Microbiol.* 133:250–260 (1982).

Lafforgue, C., Malinowski, J., and Goma, G., High yeast concentration in continuous fermentation with cell recycle obtained by tangential microfiltration, *Biotechnol. Lett.* 9:347–352 (1987).

Lee, C. W., and Chang, H. N., Kinetics of ethanol fermentations in membrane cell recycle fermentors, *Biotechnol Bioeng.* 29:1105–1112 (1987).

Lee, K. J., Lefebvre, M., Tribe, D. E., and Rogers, P. L., High productivity ethanol fermentation with *Zymomonas mobilis* using continuous cell recycle, *Biotechnol. Lett.* 2:487–492 (1980).

Mattiasson, B., and Hahn-Hägerdal, B., Microenvironmental effects on metabolic behaviour of immobilized cells. A hypothesis, *Eur. J. Appl. Microbiol. Biotechnol.* 16:52–55 (1982).

Mattiasson, B., and Larsson, M., Extractive bioconversions with emphasis on solvent production, *Biotechnol. Genetic Eng. Rev.* 3:137–174 (1985).

Mattiasson, B., Ramstorp, M., Nilsson, I., and Hahn-Hägerdal, B., Comparison of the performance of a hollow fiber microbe reactor with a reactor containing alginate entrapped cells, *Biotechnol. Lett.* 3:561–566 (1981).

McGregor, W. C., Membrane separations in biotechnology, *Bioprocess Technol.* 1 (1986).

Mehaia, M. A., and Cheryan, M., Lactic acid from acid whey permeate in a membrane recycle bioreactor, *Enzyme Microb. Technol.* 8:289–292 (1986).

Michaels, A. S., Membrane technology and biotechnology, *Desalination* 35: 329–351 (1980).

Molin, G., Nilsson, I., and StensonHolst, L., Biofilm build-up of *Pseudomonas putida* in a chemostat at different dilution rates, *Eur. J. Appl. Microbiol. Biotechnol.* 15:218–222 (1982).

Mota, M., Lafforgue, C., Strehaiano, P., and Goma, G., Fermentation coupled with microfiltration: kinetics of ethanol fermentation with cell recycle, *Bioprocess Eng.* 2:65–68 (1987).

Nishizawa, Y., Mitani, Y., Tamai, M., and Nagai, S., Ethanol production by cell recycling with hollow fibers, *J. Ferment. Technol.* 61:599–605 (1983).

Ohleyer, E., Blanch, H. W., and Wilke, C. R., Continuous production of lactic acid in a cell recycle reactor, *Appl. Biochem. Biotechnol.* 11:317–332 (1985).

Pierrot, P., Fick, M., and Engasser, J. M., Continuous acetone-butanol fermentation with high productivity by cell ultrafiltration and recycling, *Biotechnol. Lett.* 8:253–256 (1986).

Strathmann, H., Membranes and membrane processes in biotechnology, *Trends Biotechnol.* 3:112–118 (1985).

Taniguchi, M., Kotani, N., and Kobayashi, T., High concentration cultivation of *Bifidobacterium longum* in fermenter with cross-flow filtration, *Appl. Microbiol. Biotechnol.* 25:438–441 (1987a).

Taniguchi, M., Kotani, N., and Kobayashi, T., High-concentration cultivation of lactic acid bacteria in fermentor with cross flow filtration, *J. Ferment. Technol.* 65:179–184 (1987b).

Vick Roy, T. B., Blanch, H. W., and Wilke, C. R., Microbial hollow fiber reactors, *Trends Biotechnol.* 1:135–139 (1983a).

Vick Roy, T. B., Mandel, D. K., Dea, D. K., Blanch, H. W., and Wilke, C. R., The application of cell recycle to continuous fermentative lactic acid production, *Biotechnol. Lett.* 5:665–670 (1983b).

3

Liquid–Liquid Extractive Membrane Reactors

Jorge L. López and Stephen L. Matson

Sepracor Inc.
Marlborough, Massachusetts

Thomas J. Stanley

General Electric Company
Schenectady, New York

John A. Quinn

University of Pennsylvania
Philadelphia, Pennsylvania

I. INTRODUCTION

Advances in biotechnology have placed new demands on biochemical reactor engineering. This, in turn, has led to the exploration and development of several novel bioreactor configurations, many of them based on the immobilization of enzymes on solid-phase supports. Process integration—in particular the advantageous coupling of bioconversion with separation processes—is the dominant theme underlying much of the recent innovation evident in the field of bioreactor engineering.

At the same time, membrane scientists have become intrigued with the notion of "activating" membranes by incorporating reactive or catalytic species within them, the objective being to render them capable either of performing separations more selectively or of performing chemical conversions in addition to separation. An example of the former is provided by carrier-mediated or carrier-facilitated transport, in which a complexation reaction occurs between a membrane-phase carrier species and a permeant; the result can be unprecedentedly high permeability and selectivity toward that species that undergoes reversible coupling with the carrier (Smith et al., 1977; Kimura et al., 1979; Matson et al., 1983). In this chapter we discuss a class of membrane reactors in which a membrane-phase catalyst (either an enzyme or a phase-transfer catalyst) brings about net chemical conversion of reactant to product. Particular emphasis is given to membrane reactors and their bioprocessing applications in which the catalytic conversion is coupled with a liquid-liquid extraction step and one of the components of the reaction system, either the reactant or catalyst, is biological in nature.

Since the use of membranes to mediate a catalytic reaction is a relatively novel concept, a few general remarks on the rationale for and appropriateness of combining these two technologies are in order. From the viewpoint of membrane engineering, it is apparent that the activation of membranes with functional entities, such as enzymes, creates considerable opportunity to expand the role of membranes beyond performance of the simple separations for which they were originally designed and have conventionally been used. Indeed, the technology of "active membranes" (i.e., membranes with adsorptive or reactive functional groups) is now recognized as one of the more vigorous areas of synthetic membrane research and development. However, from the viewpoint of the reactor design engineer, the argument for employing membranes is not nearly as compelling at first sight. Consider, for example, the biochemical engineer charged with the design and operation of a cost-effective enzyme-catalyzed bioconversion process. It is not at all obvious that a microporous membrane is the first choice as a solid-phase support for immobilizing the biocatalyst to be used in the process. Indeed, porous membranes are much more expensive on a unit-volume basis than any of several more conventional particulate types of enzyme support (e.g., microporous particles, ion-exchange resins, or polysaccharide gel beads) that can be used in conventional reactor designs (e.g., packed or fluidized beds or stirred tanks).

Notwithstanding their relatively high costs, membrane reactors can in fact be cost-effective if the unique versatility of the membrane geometry is exploited. For example, it is possible to flow a reactant-containing solution through the pores of a catalytically active microporous membrane to efficiently transport reactants to and products away from the catalyst by convection rather than by the relatively slow process of diffusion that, of necessity, controls transport in porous catalyst particles. Furthermore, membranes compartmentalize and in so doing provide an additional interface that can exchange with a second process stream; this constitutes an additional degree of freedom without a counterpart in traditional particle-bound catalysts. Finally, membranes can be structured as multilayer laminates or composites; among other benefits, this affords the possibility of employing permselective membranes to control the fluxes of the various participants in a reaction system in a powerful way. These are the unique attributes of membranes that lead to useful and nontrivial membrane reactor function, as illustrated more specifically later.

In this chapter we discuss a family of related liquid-liquid extractive membrane reactors of potential significance in a variety of bioprocessing operations. Each operation relies on the selective partitioning of a reactant or product from one liquid phase into a second immiscible phase. The discussion follows a progression from multilayer enzyme membrane reactors based on immobilized liquid membranes, through multiphase and extractive enzyme membrane reactors in which the membrane separates immiscible aqueous and organic process streams, and finally to membrane reactors mediating the phase-transfer-catalyzed conversion of mutually insoluble reactants. Particular emphasis is given to the application of membrane reactors to the deacylation of benzylpenicillin and to the conduct of phase-transfer catalysis (PTC).

II. MULTILAYER ENZYME MEMBRANE REACTORS BASED ON LIQUID MEMBRANE EXTRACTION

The product separation and enrichment membrane reactor was originally explored by Matson and coworkers (Matson, 1979; Lopez, 1983; Matson and Quinn, 1986). This membrane reactor configuration is able to carry out catalysis, product separation from inert contaminants, and product enrichment in a single, passive device operating without an external source of energy. In Fig. 1a, a cross section of a membrane reactor composite capable of product separation and enrichment is shown. The permselective membrane is designed so that it permits reactant from a feed stream to diffuse into the catalytic film but rejects the reaction product, thus forcing its removal to take place only through a product or "sweep" stream contacting the catalytic membrane. If the reaction is essentially irreversible, the molar rate of product formation is determined by the rate at which reactant permeates into the catalytic membrane and by the kinetics of the reaction. Under conditions of constant molar conversion, the product concentration in the product stream is inversely proportional to the volumetric rate of product removal. Once the feed stream flow rate and concentrations are adjusted to give a reasonable degree of reactant removal, high degrees of product enrichment can be attained by operating the reactor at high feed-to-product stream flow rate ratios (see Fig. 1b).

By retaining the product in the sweep stream, the first step in separating it from an inert component present in the feed stream is accomplished. For this to be an effective separation, however, most of the inert component must be retained in the feed stream. In principal this can accomplished by choosing the permselective film such that it rejects the inert component as well as the product, but this imposes an additional requirement on the permselective membrane

• MEMBRANE FUNCTION

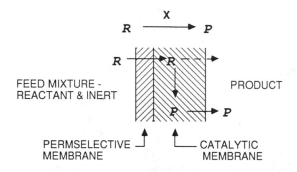

FEED MIXTURE - REACTANT & INERT

PRODUCT

PERMSELECTIVE MEMBRANE

CATALYTIC MEMBRANE

• REACTOR OPERATION

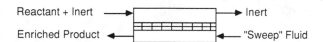

Reactant + Inert   ⟶   Inert

Enriched Product   ⟵   "Sweep" Fluid

Figure 1   Membrane bioreactor for product separation and enrichment.

that can be difficult to satisfy. Fortunately, even in the limiting case in which the permselective film is infinitely permeable to the inert component, the concentration of the inert species in the product stream cannot ordinarily be higher than its concentration in the feed stream; it follows that the molar ratio of inert component recovered in the exiting feed stream as opposed to inert component lost to the product stream approaches the feed-to-product volumetric flow rate ratio. At the same time, product stream purity is assured by the selective concentration of the reaction product relative to the inert species in that stream.

The potential benefits of the multilayer membrane reactor are associated with its ability to integrate the unit operations of catalysis, product separation, and product concentration. Not only are these three operations accomplished in a single piece of equipment, but the normally energy-consuming tasks of product separation and concentration are conducted in a particularly efficient manner that takes advantage of the availability of the free energy of the reaction.

A. ATEE Resolution

The utility of the multilayer enzyme membrane reactor in product separation and enrichment was demonstrated early by applying it to the task of amino acid resolution (Matson, 1979). Specifically, the reactor was applied to the enzymatic resolution of a racemic amino acid derivative, N-acetyl-D,L-tyrosine ethyl ester (ATEE):

$$HO\text{—}\langle\bigcirc\rangle\text{—}\underset{\underset{\text{NHCOCH}_3}{|}}{CH_2CHCOOC_2H_5} \cdot H_2O \xrightarrow[\text{pH 7.8}]{\text{ChT}} HO\text{—}\langle\bigcirc\rangle\text{—}\underset{\underset{\text{NHCOCH}_3}{|}}{CH_2CHCOO^{\ominus}} \cdot H^{\oplus} \cdot C_2H_5OH \cdot \text{D-ESTER}$$

$$\text{D,L - ESTER} \qquad\qquad\qquad\qquad\qquad\qquad\qquad\qquad \text{L-ACID} \qquad\qquad (1)$$

Since the operation of this membrane reactor has been described in some detail (Matson and Quinn, 1986), we focus the present discussion on the role of liquid membrane extraction in its operation.

Figure 2 illustrates the multilayer membrane construction employed in this reactor and depicts its operation in amino acid resolution. The feed stream consists of an aqueous solution of the racemic amino acid ester (DL-ATEE), which flows by the surface of the permselective membrane in the two-layer membrane sandwich. The opposite surface of the laminate (i.e., the interface with the enzyme-activated membrane) is swept with a second aqueous stream that contains a buffer and serves to carry off the separated and concentrated reaction product. In operation, the L-selective enzyme (immobilized chymotrypsin) in the catalytic membrane converts L-ester (but not D-ester to the corresponding L-acid; the result is separation and enrichment of L-acid in the product stream and recovery of the unreactive D-ester in the exiting feed stream.

The permselective membrane used to control the fluxes of reactant and product was an immobilized liquid membrane (ILM) (Ward, 1972). An ILM consists of a water-immiscible organic solvent for the permeant that is immobilized by capillarity in the pores of a microporous, hydrophobic support membrane. ILMs prepared in this fashion combine the desired permeation properties of liquids (e.g., high permeant solubilities and diffusivities compared to their corresponding values in polymeric membrane materials) with the mechanical and geometrical characteristics of the microporous support film. In effect, ILMs permit one to conduct solute separations based on liquid-liquid extraction in a stable membrane configuration, with extraction occurring at one membrane interface and back-extraction (stripping) occurring at the other. Such membrane-based liquid-liquid extractions are particularly useful for providing the high reactant/product selectivity desired in multilayer membrane reactor applications.

• **MEMBRANE FUNCTION**

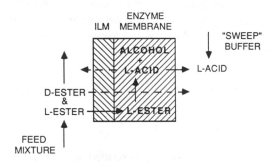

• **REACTOR OPERATION**

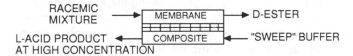

Figure 2 Multilayer membrane bioreactor applied to amino acid resolution.

Our ILMs were prepared by impregnating a highly microporous Teflon membrane (Goretex PTFE film) with 1-decanol, which is a good solvent for the amino acid ester of interest here. Such ILMs met or exceeded our requirements in each of the three performance areas important to membrane bioreactor operation: reactant permeability, reactant/product permselectivity, and stability. The permeability of these films to the reactant L-ester was very high (3×10^{-4} cm/s), and reactant/product selectivity (i.e., the ratio of L-ester to L-acid permeabilities) exceeded 15,000. Finally, these ILMs proved stable for up to 6 weeks of continuous operation, which was more than sufficient for experimental purposes.

When this immobilized liquid membrane was combined with an immobilized enzyme membrane and the resulting sandwich was fed a racemic amino acid ester mixture, efficient separation of L- and D-enantionmers was achieved, as shown in Fig. 3. The product stream consisted primarily of separated and 10-fold enriched L-acid, with the bulk of the inert D-isomer recovered in the exiting feed stream at modest optical purity. This system served as the initial demonstration of product separation and enrichment and provided an early indication of the potential significance of membrane reactor technology to biocatalysis and, more specifically, the optical resolution of chiral compounds.

B. Deacylation of Benzylpenicillin

The potential of multilayer enzyme membrane reactors based on immobilized liquid membrane extraction for the integration of multistep bioprocess operations has also been demonstrated by its application to the deacylation of benzylpenicillin (Lopez, 1983). On one hand, the enzymatic conversion of fermentation-derived penicillins to 6-aminopenicillanic acid (6-APA) is a process of considerable industrial importance. Additionally, however, this particular application was attractive in that it permitted us to explore a new liquid-liquid extraction chemistry, in particular, that of ion-pair extraction, in the context of membrane reactor bioprocessing.

The large-scale production of semisynthetic penicillins is based on the deacylation of benzylpenicillin (penicillin G) pr phenoxymethylpenicillin (penicillin V) to produce 6-aminopenicillanic acid (6-APA), a β-lactam thiazolidine ring system that is the nucleus of all penicillin antibiotics (Rhodes and Fletcher, 1966). This transformation can be accomplished through enzymatic or chemical processes (Carrington, 1971). The enzymatic route makes use of the enzyme penicillin acylase (PA), which is found in a wide variety of microorganisms (Vandamme and Voets, 1974). The reaction stoichiometry for this transformation is

In recent years a large number of enzyme isolation, purification, and immobilization techniques have been developed, making possible the use of highly purified enzyme preparations for extended periods of time. Several of these techniques have already been applied to the isolation of penicillin acylase and its use in continuous reactor systems for the production of 6-APA (Chiang and Bennet, 1967; Carleysmith and Lilly, 1979; Park et al., 1982). Because biochemical substrates, such as benzylpenicillin, are rarely found in a pure and/or

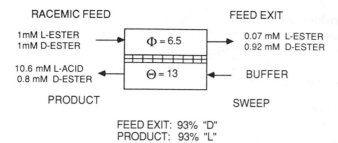

RACEMIC FEED   FEED EXIT

1mM L-ESTER   0.07 mM L-ESTER
1mM D-ESTER $\Phi = 6.5$ 0.92 mM D-ESTER

10.6 mM L-ACID   BUFFER
0.8 mM D-ESTER $\Theta = 13$

PRODUCT   SWEEP

FEED EXIT: 93% "D"
PRODUCT: 93% "L"

Figure 3 Amino acid (BTEE) resolution in a multilayer membrane bioreactor.

concentrated form, costly separation and enrichment operations must generally be carried out before and after the reaction step. The advantages inherent in the liquid-liquid membrane reactors described here offer a unique opportunity for the manufacture of semisynthetic penicillins.

1. Carrier-Mediated Transport

The membrane reactor configuration used in this work was similar to that used by Matson in his original demonstration. The permselective membrane was the same type used by Matson, an immobilized liquid membrane. This type of membrane consists of a microporous, hydrophobic support matrix impregnated with a suitable organic solvent confined to the pores by capillary action. Benzylpenicillin has a carboxylic acid group with a pK of approximately 2.75, and therefore at neutral pH it has a negative charge, making its solubility in most organic solvents extremely low.

To facilitate the permeation of benzylpenicillin through an ILM it is necessary to neutralize the negative charge. Several methods of achieving this were studied. The first involved partial acidification of the penicillin solution to drive the acid equilibrium toward its uncharged state. However, early in this investigation it was realized that this approach would have serious limitations since the pH required to obtain an acceptable partition coefficient (~ 3) between the organic solvent in the ILM and the aqueous phase was such that substantial degradation of benzylpenicillin would occur.

Two other methods can be employed to promote the partition of benzylpenicillin into organic solvents. These are based on the use of adduct- and ion-pair-forming agents to extract organic acids from aqueous solutions into non-aqueous solvents (Modin, 1971; Modin and Schroder-Nielsen, 1971; Schill, 1974). The adduct-forming reaction can be expressed as

$$R'COO^-_{aq} + H^+ + R_3 N_{org} < - - - - > R'COOH \cdot R_3 N_{org} \tag{3}$$

and the ion pair reaction is

$$R'COO^-_{aq} + R_4 N^+_{aq} < - - - - > R'COOR_4 N_{org} \tag{4}$$

After an experimental screening of these two extraction procedures, the ion-pair–forming reaction was chosen to transport benzylpenicillin through an ILM. The ion-pair–forming agent was the same used by Modin and Schroeder-Nielsen, tetrabutylammonium (TBA) (1971). The extraction constant for the reaction is defined as

$$E = \frac{[R'COOR_4N]_{org}}{[R'COO^-]_{aq}[R_4N^+]_{aq}} \tag{5}$$

This constant was experimentally determined for both benzylpenicillin and 6-APA. The value for penicillin was found to be 60 M^{-1}. This means that a TBA concentration of 0.05 M in the aqueous phase is needed for the partition coefficient of benzylpenicillin to be 3.0. The extraction constant determined for 6-APA was 1.2 M^{-1}.

The use of tetrabutylammonium to promote the transport of benzylpenicillin through an immobilized liquid membrane is an example of carrier-mediated transport. The versatility and usefulness of this mode of transport has been gaining increasing attention in the last few years (Schultz, 1977; Way et al., 1982). However, other permeation modes and/or membranes may also be used as long as the requirement of high reactant permeability and low product permeability are met.

2. Membrane Reactor Models

Modeling of the type of membrane reactor described here is a complex problem, and the reader is referred to the work of Matson and Lopez for a more detailed description and results. Nevertheless, membrane reactor performance can be predicted qualitatively and to a large extent quantitatively by solving analytically or numerically the equations that describe such a reactor using kinetic and transport parameters determined from independent measurements.

Several dimensionless groups are obtained when the equations that describe these membrane reactors are nondimensionalized. Among the most important are the following. Reactor space-time:

$$\Phi = \frac{D^S P^E S Q^\circ A_m}{1q^f}$$

Flow rate ratio:

$$\Theta = \frac{q^f}{q^p}$$

Thiele modulus:

$$\phi = \delta \left(\frac{k_0}{D_{eff}}\right)^{1/2}$$

Selectivity:

$$Pp = \frac{D^PpE^P}{D^SpE^S}$$

In Fig. 4 the effect of reactor space-time on reactant removal from the feed stream is shown for several values of Thiele modulus. Note that since reactor space-time is proportional to the feed carrier concentration, reactant removal can be increased by adding more carrier (ion-pair–forming agent) to the feed solution.

Figure 5 shows the effect of selectivity on the product enrichment factor (PEF), defined as the ratio of product concentration in the product stream to the reactant concentration entering the reactor as a function of the flow rate ratio. The limiting product enrichment factor is obtained by multiplying the

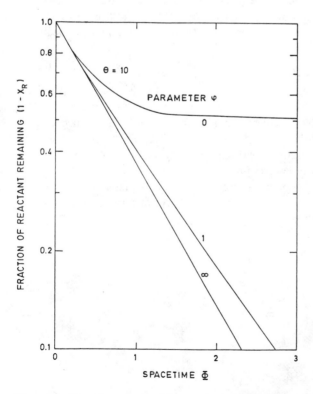

Figure 4   Reactant removal in a cocurrent plug flow reactor high flow rate ratio.

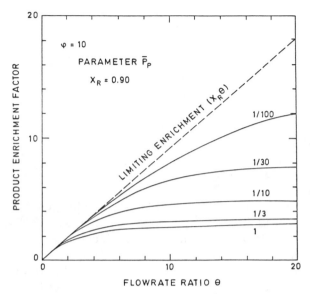

Figure 5 Effect of selectivity and flow ratio on product enrichment factor.

flow rate ratio by the reactant removal. For any selectivity greater than zero, or a modest Thiele modulus, the enrichment is less than the limiting PEF. At fixed flow rate ratio, the enrichment factor varies inversely with the selectivity. For a given Thiele modulus, the maximum PEF is obtained when the selectivity is equal to zero.

3. Experimental Demonstration

Benzylpenicillin sodium salt (1670 IU/mg) and 6-aminopenicillanic acid were purchased from Sigma Chemical Co. Tetrabutylammonium hydrogen sulfate and n-decanol were obtained from Aldrich Chemical Co. The Goretex membranes were obtained from Gore Associates (Maryland) and the cellulose membranes from Millipore Co. Penicillin acylase obtained from strains of *Bacillus megaterium* was supplied by Wyeth Laboratories (West Chester, PA).

Determination of 6-APA concentrations was done using the p-dimethylaminobenzaldehyde (PDAB) method described by Balasingham et al. (1972). Benzylpenicillin concentrations were determined using Boxer and Everett's (1949) hydroxylamine method for total penicillins. In the presence of 6-APA, the concentration of benzylpenicillin was determined by subtracting the concentration of 6-APA found with the PDAB method from the total penicillin concentration determined with the hydroxylamine method.

The following solutions were used for the feed and product streams:

Feed
    0.005 M benzylpenicillin
    0.05 M tetrabutylammonium
    0.05 M $K_2 HPO_4$
    pH 6.5
Product
    0.1 M $H_3 BO_4$
    0.05 M $K_2 HPO_4$
    pH 8.5

All buffer solutions were saturated with decanol prior to their use to prevent loss of decanol from the ILM.

The catalytic membrane was prepared as follows. Penicillin acylase (from 8 to 5 μmol/min·mg in 2% benzylpenicillin at 25°C) was covalently attached to Millipore-type MF filters (HAWP, 0.45 μm) using the cyanogen bromide technique (Axen et al., 1967). Typically, two 9 X 5 inch membrane strips weighing approximately 2 g each were activated with 8 g of cyanogen bromide at pH 11.0. The activation was carried out in 1500 ml water, and the pH was maintained by the addition of 6 M NaOH. After the caustic consumption ceased, the membranes were rinsed in cold water and then left to soak overnight in 2 L of a 1 mg/ml enzyme solution in 1 M $NaHCO_3$. Membrane activity was determined by placing a membrane disk in a filter holder and passing a 2% benzylpenicillin solution in 0.1 M TRIS buffer, pH 8.5, through the membrane. 6-APA concentrations were measured in the effluent solution. The activities obtained were normally between 0.1 and 0.2 μmol of 6-APA formed per min per cm^2 of membrane at 25°C (11–22 mM/s).

Immobilized liquid membranes were prepared by carefully mounting Goretex polytetrafluorethylene membranes (1 μm nominal pore diameter, 85% porosity, 44 μm thickness) in a previously cut plastic frame and stapling the membrane to the frame. Once mounted, the membranes were placed in a decanol-filled tray. Air trapped within the pores upon immersion usually disappeared in 1 or 2 days. Prior to assembly of the membrane, composite excess solvent on the membrane surface was removed by blotting the film between tissues.

A membrane reactor similar to that described by Matson was used in this study. Figure 6 shows a typical schematic diagram. The active area of the reactor was 200 cm^2, and provisions were made to keep the protruding edges of the membranes immersed with decanol to prevent wicking. Feed and product solutions were delivered by calibrated Sage infusion-withdrawal syringe pumps. Prefilters were placed between the pumps and reactor inlets.

Reactor start-up normally involved starting the feed and product pumps and making sure that the separator screens which were glued to the plexiglass at six locations, were completely wet with solution. The catalytic film was placed

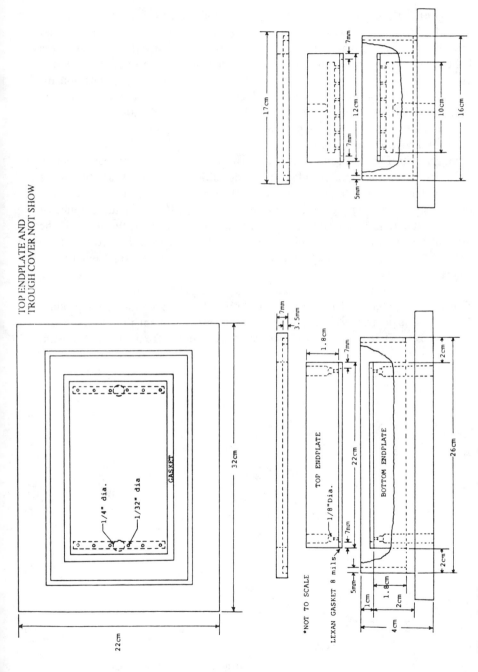

Figure 6 Membrane reactor schematic.

on top of the bottom plate, and then the immobilized liquid membrane was laid over it. The top plate was positioned on top of the membrane composite, and the plastic frame on which the ILM was mounted was cut out. The plates were held together in position by 10 C clamps that applied pressure on the gasket area. Once the reactor was assembled, it was placed in a vertical position for some time to remove any trapped bubbles from the compartments. A trough built around the bottom plate of the reactor was then filled with n-decanol until the outer edges of the membranes were completely immersed in solvent.

A summary of the results obtained in this study is presented in Figs. 7 through 9. In Fig. 7, reactant removal is shown as a function of reactor space-time. All the parameters included in the definition of reactor space-time were measured experimentally; only the feed flow rate was varied to obtain the desired space-time. The product flow rate was 0.1 ml/min. In the absence of mass-transfer resistance in the feed and product compartment, the slope of the line shown in this figure should be approximately equal to $1/\ln 10$. That the slope is less, $0.41/\ln 10$, suggests the presence of significant intrachannel mass-transfer resistance.

In Fig. 8, the effect of flow rate ratio on the product enrichment factor is shown. The product enrichment factor is defined here as the ratio of the product concentration in the feed stream. It is convenient to present the results in this form because it is possible to obtain reactant removal from the feed stream,

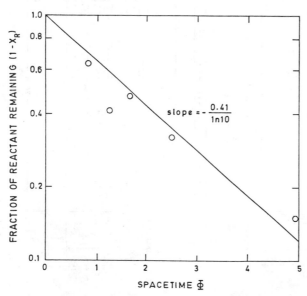

Figure 7 Reactant removal as a function of membrane reactor space-time.

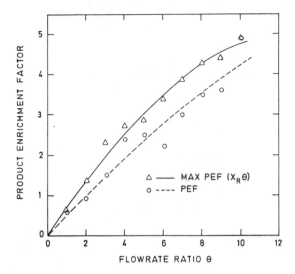

Figure 8 Effect of flow rate ratio on product enrichment factor.

reactant conversion, and the fraction of unconverted reactant in the product stream from this type of plot. From material balance considerations, and assuming no product leakage into the feed stream, the following relations apply:

$$\text{Reactant removal} = \frac{\text{PEF}_{max}}{\theta}$$

$$\text{Reactant conversion} = \frac{\text{PEF}}{\theta}$$

$$\text{Reactant in product} = 1 - \frac{\text{PEF}}{\text{PEF}_{max}}$$

In this study, as in the work by Matson, no significant amount of product was detected in the feed stream. For this experimental run the product flow rate was varied randomly while keeping the feed flow rate at 1 ml/min. It is important to note here that an uncontrolled variable in these experiments was the loss of enzymatic activity throughout the duration of the experimental runs, which sometimes lasted for as long as 8 days. Nevertheless, the qualitative behavior of the experimental results agrees fairly well with the behavior predicted from theory (see Fig. 5). A typical set of the reactor operating conditions is shown in Fig. 9.

It is evident from the results of this study that the use of liquid-liquid-enzyme membrane reactors for the deacylation of benzylpenicillin is an intriguing

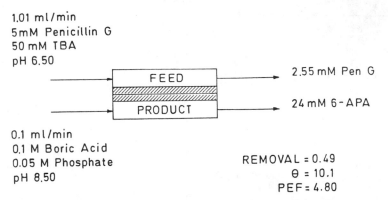

1.01 ml/min
5mM Penicillin G
50 mM TBA
pH 6.50

2.55 mM Pen G

24 mM 6-APA

0.1 ml/min
0.1 M Boric Acid
0.05 M Phosphate
pH 8.50

REMOVAL = 0.49
θ = 10.1
PEF = 4.80

Figure 9 Typical membrane reactor operating conditions.

alternative to conventional continuous reactor designs. The complexity of this system, however, diminishes to a certain degree the attractiveness of this reactor concept.

III. MULTIPHASE AND EXTRACTIVE ENZYME MEMBRANE REACTORS

Considerable progress has been made since the previously described work on multilayer membrane reactors based on liquid membrane extraction. In particular, the capabilities of membrane reactors have been extended beyond process integration to the solution of several other important bioprocessing problems, such as the bioconversion of sparingly soluble substrates and improvement of product yields in product-inhibited reaction systems. Product separation and enrichment have been retained as key aspects of membrane bioreactor operation, even while the configuration of these membrane reactors has been simplified dramatically. Whereas the earliest membrane reactors invoked multiple layers of flat-sheet membrane, single-layer hollow-fiber membranes have been used more recently, without sacrificing reactor performance.

The selective partitioning of solutes (e.g., reactants and products) between immiscible liquid phases remains the principal basis for managing the reaction participants in these more recent multiphase and extractive membrane reactors. The difference is that whereas earlier multilayer membrane reactors employed extraction solvents in the configuration of immobilized liquid membranes, and both the feed and product streams consisted of aqueous solutions, the current generation of single-layer enzyme membrane reactors finds application in mediating reaction and separation processes in which reactants and products are found in immiscible aqueous- and organic-phase process streams. An extraction process is thus inherent to this generation of membrane reactors. When the reactant is initially present in a water-immiscible organic process stream and the direction of reactant extraction is from the organic to the aqueous phase, we refer to the reactor as a "multiphase" enzyme membrane reactor. When the feed is aqueous and a product is extracted into an organic process stream, the reactor is termed "extractive." Both are described briefly here.

A. Multiphase Enzyme Membrane Reactors

1. Bioconversions of Sparingly Soluble Substrates

Many chiral resolution opportunities involve organic chemicals that are sparingly soluble in water. This presents special problems when enzyme-based resolution methods are considered. Clearly, one problem is finding an enzyme that has activity and the desired stereoselectivity on what usually is an unnatural substrate. Enzymes have been identified, however, that are particularly useful in the production or resolution of carboxylic acids, esters, and alcohols. A remaining problem is to perfect reactor process designs that make enzyme-based schemes economically feasible on a commercial scale.

The problem with existing dispersed-phase reactor systems can be appreciated by close examination of the mass-transfer and reaction steps. Consider the generalized case in which it is desired to produce a resolved carboxylic acid. A preferred enzymatic approach is to use a hydrolytic enzyme, such as a lipase or esterase, to stereoselectively convert an ester to its respective acid and alcohol:

$$R'CO_2R'' + H_2O \longrightarrow R'CO_2H + R''OH \qquad (6)$$

The acyl moiety R' in many of the chiral compounds of commercial interest contains substituted benzyl, naphthyl, and/or aryloxy functionalities, which impart water insolubility to the ester. The resulting acids, however, exhibit significant water solubility since most often the optimum pH for the reaction is above their pK_a value.

Such reactions are conventionally carried out in a multiphase stirred-tank reactor or in a catalytic packed-bed configuration. In either case, reactor productivity can be poor. The nature of the productivity limitation is shown schematically in Fig. 10. The organic substrate is typically dispersed in a continuous aqueous phase, either "neat" or dissolved in an appropriate immiscible organic solvent. The enzyme catalyst is immobilized on, or contained within, a porous support of some sort. For the reaction to proceed, the substrate must first partition into the aqueous phase and be transported to the supported catalyst. However, its low water solubility limits the diffusive flux of substrate to the catalyst and hence causes reactor productivity to suffer. Additional problems associated with dispersed-phase reaction systems relate to difficulties in phase separation and product recovery.

Figure 11 shows a cross-sectional view of the enzyme membrane in a multiphase membrane reactor designed to address many of these limitations of

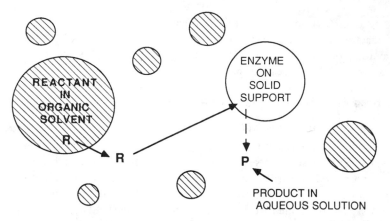

Figure 10 Conventional dispersed-phase reaction system.

dispersed-phase reactors. A hydrophilic, microporous membrane, suitably activated by incorporation of an enzyme within its pores, is disposed at the interface between aqueous and organic process streams. In operation, reactant partitions from the organic-phase feed stream into the water-wet enzymatic membrane, where it is subsequently converted to the water-soluble product. The product then diffuses out into the aqueous process stream.

The membrane in a multiphase membrane reactor thus serves three functions. First, it provides high surface area contact between the two immiscible process streams on either side of it. The membrane serves to separate the bulk phases, thus avoiding the need to disperse one phase within the other. Finally the enzyme-activated membrane also functions as an interfacial catalyst, placing the supported enzyme in direct contact with the organic phase containing the reactant. In this manner, it is possible to minimize the diffusive limitations associated with the intervening bulk aqueous phase characteristic of dispersed-phase

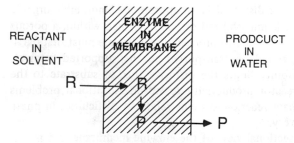

Figure 11 Schematic of multiphase enzyme membrane reactor operation.

bioreactors. Hollow-fiber membrane modules are particularly attractive as multiphase membrane reactors (see Fig. 12).

2. BTEE Resolution

The application of a multiphase membrane reactor to the enzymatic resolution of racemic amino acids can be illustrated using a biochemical system quite analogous to that described for multilayer membrane reactors. In this case, we choose to resolve a racemic mixture of N-benzoyl tyrosine ethyl ester or DL–BTEE, as opposed to the related N-acetyl compound used in the previous studies. The water solubility of DL–BTEE is approximately 0.2 mM.

The 1.0 m^2 multiphase membrane reactor employed for BTEE resolution consisted of polyacrylonitrile-based hollow-fiber membranes activated by the incorporation of 4 g of chymotrypsin. This enzyme exhibits essentially perfect stereoselectivity for the hydrolysis of L-BTEE to the corresponding acid, N-benzoyl-L-tyrosine of L-BT. In this particular case, the enzyme was first adsorbed to the micropore walls; subsequently, the adsorbed protein layer was stabilized by covalently cross-linking it with a 2.5% solution of glutaraldehyde. The poorly water-soluble racemic ester was then fed to the membrane reactor as a 30 mM solution in 1.1 L of octanol; the opposite surface of the membrane (see Fig. 13) was contacted with 0.2 L of an aqueous phosphate buffer (pH 7.8) to maintain reaction pH and provide a reservoir for the water-soluble reaction product. Both the organic and aqueous phases were then recycled to the multiphase membrane reactor, with the batchwise reaction proceeding essentially to completion overnight.

Table 1 summarizes concentrations and quantities of each of the species present in the reaction system before and after the run; both organic- and

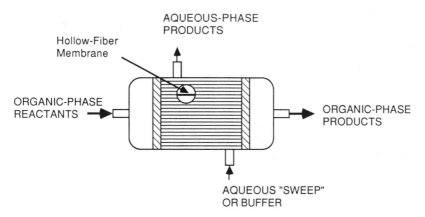

Figure 12 Hollow-fiber multiphase membrane reactor.

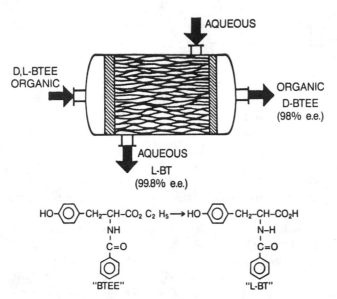

Figure 13 Resolution of DL–BTEE in a multiphase membrane bioreactor.

aqueous-phase concentrations are tabulated. Nearly all the relatively water-soluble reaction product, the L–BT acid, was found in the aqueous process stream, where it was enriched approximately fivefold over the initial concentration of L-ester in the feed stream. In contrast, the inert D–BTEE ester was recovered in high yield from the organic stream, A final accounting of the concentrations of the various enantiomers yielded enantiomeric excess values of 98% (i.e., [D − L]/[D + L]) and 99.8% (i.e., [L − D]/[L + D]) in the organic and aqueous phases, respectively.

This example serves to illustrate the ability of single-layer, hollow-fiber–based multilayer predecessors, to separate and enrich reaction products in systems involving the organic-phase feed of water-insoluble reactants. Moreover, it highlights an important area for commercial application of the technology, namely, the enzymatic resolution of sparingly water-soluble chiral pharmaceuticals and agricultural chemicals. Significantly, in both the multilayer and multiphase membrane reactors, selective partitioning of reactants and products between immiscible organic and aqueous phases provides the basis for controlling the disposition of the reaction participants.

3. Process Integration in 6-APA Production

Another promising area for the commercial application of multiphase membrane reactor technology in bioprocessing is in the enzymatic deacylation of the penicillins, the same application as described in connection with multilayer enzyme membrane reactors.

Table 1 Enzymatic Resolution of DL–BTEE in a Multiphase Membrane Bioreactor[a]

Species	Stream	Initial		Final	
		Concentration (mN)	mmol	Concentration (mM)	mmol
D–BTEE	Organic	14.9	16.8	14.9	16.79
L–BTEE	Organic	14.9	16.8	0.0	0.0
L–BT	Organic	0.0	0.0	0.12	0.14
D–BTEE	Aqueous	0.0	0.0	0.07	0.01
L–BTEE	Aqueous	0.0	0.0	0.0	0.0
L–BT	Aqueous	0.0	0.0	80.0	16.7

[a]Membrane: 4 g chymotrypsin on 1 m^2 PAN hollow-fiber module. Organic phase: 1.13 L 1-octanol. Aqueous phase: 0.21 L phosphate buffer, pH 7.8. Organic-phase enantiomeric excess, 98%; aqueous-phase enantiomeric excess, 99.8%.

Figure 14 shows a general process diagram for the manufacture of 6-APA. The process shown is in reality two independent processes. The first is the manufacture of pure crystalline penicillin G or V. In general, the fermentation step accounts for 50-60% of the final cost of the pure penicillin, the balance of this cost corresponding to the extensive isolation and purification steps required to meet pharmaceutical standards. Depending on the demand for semisynthetic penicillins and other factors, a portion of this pure penicillin is subsequently fed to an enzymatic reactor to produce 6-APA. Because the raw material to this reactor is pure penicillin its contribution to the final cost of 6-APA is substantial, about 75% of the total.

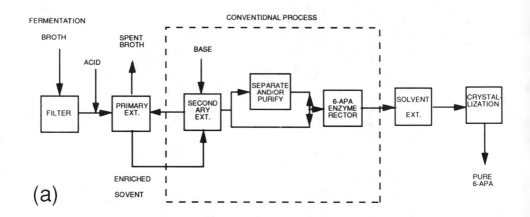

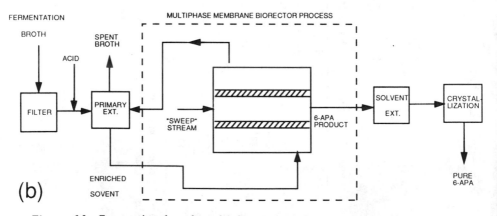

Figure 14 Conventional and multiphase membrane reactor processes for the manufacture of 6-APA.

An attractive approach currently being investigated is integration of the enzymatic reaction with the penicillin purification process. In one configuration, the penicillin-enriched organic stream coming from the primary extraction is fed to a multiphase membrane reactor. The organic solvent contacts a membrane in which the enzyme has been immobilized. The penicillin acid is extracted into the membrane, where it is enzymatically converted to 6-APA. The reaction product exits the membrane via an aqeuous stream on the other side of the membrane. The advantages of this approach can best be seen in Fig. 14. This novel approach to the deacylation of penicillins has already been demonstrated experimentally.

B. Extractive Membrane Reactors

Yet another variation on the theme of enzyme-activated membrane reactors used to mediate bioconversions involving immiscible process streams is provided by the so-called extractive membrane reactors. Here, the objective is increased product yield in biochemical reaction systems that are inhibited by reaction products, either because the enzyme catalyst is product inhibited or because the reaction is reversible. An effective reactor design strategy in such situations is to remove the inhibitory reaction product as expediently as possible, and this can be accomplished by extracting it into a separate phase. Figure 15 shows an extractive membrane reactor useful when water-soluble reactants (e.g., carboxylic acids and alcohols) are to be converted enzymatically into organic soluble reaction products (e.g., esters) via esterification reactions, which are thermodynamically unfavorable in aqueous media. Extractive reaction is, of course, a well-established technique. However, we were particularly intrigued with the convenience and efficiency of conducting extractive reactions in membrane

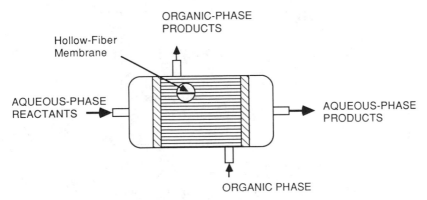

Figure 15 Hollow-fiber extractive membrane reactor.

systems in which the membrane serves the multiple roles of phase contractor, phase separator, and interfacial catalyst.

BT Esterification

For demonstration purposes, we chose to use a chymotrypsin-activated extractive membrane bioreactor to esterify the amino acid derivative N-benzoyl-L-tyrosine (L–BT) with ethanol, simply reversing the BTEE hydrolysis reaction used to illustrate the operation of the corresponding multiphase membrane reactor. This esterification reaction is highly unfavorable under the aqueous conditions in which enzymes are normally employed, proceeding to a very low fractional conversion of approximately 0.0014 at the aqueous-phase concentrations employed in our extractive membrane reactor study.

However, by selectively removing the inhibitory product into an organic extractant (octanol, in this case), significantly higher degrees of acid conversion to ester can be realized. Figure 16 shows the fractional conversion of L–BT to L–BTEE observed as a function of time in an extractive membrane reactor run batchwise at two different organic-to-aqueous phase ratios. The highest fractional conversion observed (0.35) represents a significant improvement over that possible in a homogeneous reaction system, and the extractive membrane reactor provides a particularly convenient and efficient reactor configuration.

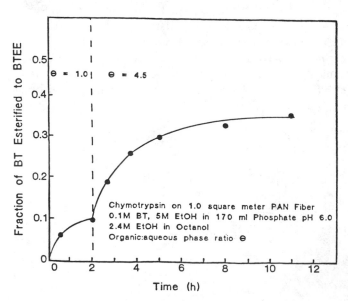

Figure 16 BT esterification in an extractive enzyme membrane reactor.

IV. PHASE-TRANSFER CATALYSIS IN A MEMBRANE REACTOR

Our initial focus in membrane reactor development was limited to enzymatic catalysis since membranes capable of withstanding the relatively mild conditions associated with biocatalysis were readily available, as were high-value-added applications. However, we now appreciate that many of the same considerations apply to synthetic (as opposed to biological) homogeneous catalysis and that the merits of membranes as fields for chemical reaction are not limited to biocatalysis. The extension of membrane reactor technology to nonbiological catalysis is made even more natural when one realizes that multiphase catalytic reaction systems occur frequently in industrial chemistry, with phase-transfer-catalyzed chemistries providing but one example. Within the present context of extractive reaction systems of pertinence to bioprocessing, it is also significant that phase-transfer catalysis (PTC) can be employed in the production of semisynthetic penicillins.

The most convenient synthetic pathway to high-value-added biological products frequently involves a combination of fermentation (or enzymatic) conversions and "conventional" chemical conversions. The most significant impact of phase-transfer catalysis in the pharmaceutical industry is the modification of natural penicillins. We illustrate the importance of PTC in bioprocessing with this example and emphasize that PTC has been enthusiastically adopted for the final tailoring of a number of other biomolecules, including amino acids and a host of other pharmacological intermediates (Freedman, 1986).

The fermentation product benzylpenicillin is transformed into semisynthetic penicillins by both enzymatic and chemical means. A clever example involves esterification of the carboxylic acid group on ampicillin, resulting in an organophilic molecule (e.g., talampicillin and bacampicillin). The activity of the penicillin is recovered as the ester group slowly hydrolyzes in the body, thereby achieving a controlled-release effect. The most effective esterifying agents are alkyl halides. Because alkyl halides rapidly hydrolyze in water, the reaction can be conducted with much higher selectivity in an organic environment. A phase-transfer catalyst is used to extract the anionic penicillin into the organic phase, where esterification takes place. The catalyst then returns to the aqueous phase, with the displaced halide ion leaving the new, organophilic penicillin in the organic phase (Reuben and Sjoberg, 1981).

The term "phase-transfer catalysis" is commonly applied to a technique used to promote reactions between aqueous- and organic-phase reactants that normally do not proceed because of their mutual insolubility. As usually employed, a positively charged phase-transfer agent is added to a two-phase mixture to extract an aqueous-phase anion into the organic phase as an uncharged ion pair;

in this way the reactants are brought into contact in the organic phase in reactive form. The traditional approach to achieving mutual solubility is to add a cosolvent to the two-phase mixture that dissolves both reactants. However, cosolvents that hydrogen-bond (e.g., ethanol) tend to solvate the anions so strongly that the overall reaction rate remains prohibitively slow. Dipolar aprotic solvents (e.g., acetonitrile or dimethylformamide) work well, but these solvents are expensive and difficult to separate from the reaction mixture. Phase-transfer catalysis, introduced by Starks (1971), eliminates the difficulties encountered when using cosolvents and is today the technique of choice for conducting this class of reactions. Many reviews of PTC have appeared in the recent literature (Dehmlow, 1977; Gokel and Weber, 1978; Brandstrom, 1977; Montanari et al., 1982; Ford and Tomoi, 1984). Also, three monographs on the subject are now available (Starks and Liotta, 1978; Weber and Gokel, 1977; Dehmlow and Dehmlow, 1980).

Over the last 10 years, PTC has become a standard tool used by chemists for laboratory synthesis, and very recently, the advantages of PTC for industrial-scale processing have been recognized; today, many specialty chemicals are produced via this technique (*Chem. Week*, 1983; Reuben and Sjoberg, 1981; Stinson, 1986; Freedman, 1986).

PTC plays an increasingly important role in biotechnology. The most convenient synthetic pathway to a high-value-added biological molecule frequently involves a combination of fermentation (or enzymatic) conversions and conventional chemical conversions. An important example is found in the production of synthetic penicillins. The fermentation product benzylpenicillin, the precursor molecule for a large variety of synthetic penicillins, is transformed into synthetic penicillins by chemical conversion to optimize pharmacological potency. PTC has been enthusiastically adopted by the pharmaceutical industry for the production of synthetic penicillins and for the final tailoring of a number of other biomolecules (Freedman, 1986).

The mechanism of phase-transfer catalysis for a simple displacement reaction is outlined in Fig. 17. The catalyst Q^+ extracts the aqueous reactant anion Y^- into the organic phase as an uncharged ion pair QY. This ion pair then reacts in the organic bulk, with the organic-phase reactant RX forming the desired product RY and a product ion pair QX. The product ion pair moves to the phase interface, where it dissociates. Reaction conditions are most favorable when the extraction of reactant Y^- with Q^+ as ion pair QY is favorable to drive the active catalyst into the organic phase, and the extraction of product anion X^- with Q^+ is unfavorable so that the spent catalyst QX readily rejects coproduct to the aqueous phase. It is then free to extract fresh reactant and is thus "regenerated." Under these conditions, much less than stoichiometric quantities of the catalyst can be used because the catalyst can shuttle back and forth between phases, repeatedly bringing fresh reactant into the organic phase.

$$Q^+ \; + \; Y^- \; \longleftrightarrow \; QY \qquad\qquad E^y = \dfrac{(QY)}{(Q^+)(Y^-)}$$

$$QY \; + \; RX \; \longrightarrow \; QX \; + \; RY \qquad\qquad r = k^f(QY)(RX)$$

$$QX \; \longleftrightarrow \; Q^+ \; + \; X^- \qquad\qquad 1/E^x = \dfrac{(Q^+)(X^-)}{(QX)}$$

Organic

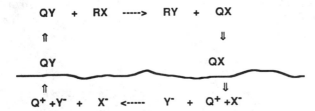

Aqueous

Figure 17 Mechanism of phase-transfer catalysis. The ion-pair formation and dissociation steps are fast compared to the bulk organic-phase displacement reaction.

Evidence supporting this mechanism of PTC has been provided by many workers (Starks and Owens, 1973; Landini et al., 1975, 1977, 1978; Gordon and Kutina, 1977). The role of mass transfer in PTC has been studied by Evans and Palmer (1981), and Evans (1983), Melville (1983), and Melville and Goddard (1985).

There are two major classes of phase-transfer catalysts: quaternary phosphonium or ammonium salts (in general, onium salts) and chelating agents (inclusion compounds), such as crown ethers and cryptands. For liquid-liquid systems, onium salts are usually as effective as the chelating agents and considerably less expensive. The most effective onium salts are relatively large so that they are highly hydrophobic and strongly partitioned into the organic phase. Also, the best catalysts are symmetrical to maximize charge separation in the ion pair and thereby to maximize reactivity with the organic-phase substrate (Herriott and Picker, 1975; Brandstrom, 1977). Although an onium salt is the catalyst of choice for liquid-liquid systems, a study by Yee et al. (1986) indicates that crowns can produce much faster rates in solid-liquid systems.

On an industrial scale, phase-transfer reactions are carried out by dispersing the aqueous and organic phases in mixer-settler (mixer-separator) contacting equipment and then separating the dispersed phases downstream of the reactor for product and catalyst recovery (Fig. 18). This separation step can be difficult if the catalyst is surface active, if there are surface-active impurities in the mixture, or if the phase density difference is small. A membrane reactor (Fig. 18) can be employed for carrying out such reactions in which the two reactant streams are brought into contact by flowing them along opposite sides of a porous membrane (Stanley, 1986; Stanley and Quinn, 1987); the role of the membrane is to localize the aqueous/organic interface and to render it stable over a wide range of operating conditions (Kiani et al., 1984: Qi and Cussler, 1985a,b). The emulsification problems that frequently plague conventional dispersed-phase reactors are avoided in the membrane reactor because no dispersion is formed. Other advantages of this membrane device are that it is passive and easily scaled up. Finally, the membranes required need not show selectivity because the membrane function is simply to stabilize the phase interface at a fixed position.

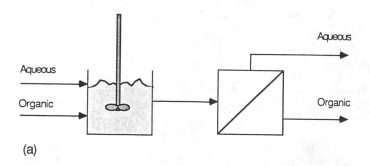

(a)

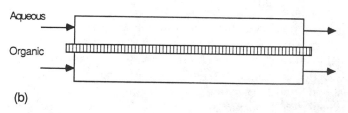

(b)

Figure 18 Schematic of industrial processing alternatives for phase-transfer catalysis: (a) dispersed-phase reactor; (b) membrane reactor.

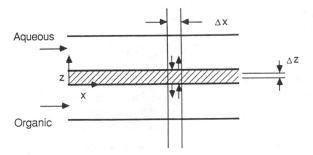

Figure 19 Diagram of a differential length of the membrane reactor.

A. Theoretical Model

In this section we present a mathematical model of an experimental flat-sheet membrane reactor operated in continuous, cocurrent mode. The goal of this modeling work is to identify the dimensionless groups governing reactor performance, to examine the sensitivity of the reactor performance to the most important of these dimensionless groups, and to provide a basis for comparison of experiment and theory.

The membrane plays a particularly simple role in this system, that of localizing and stabilizing the phase interface. Membranes useful in the conduct of phase-transfer–catalyzed reactions are generally finely porous, with one of the two liquid phases filling the void volume of the porous membrane and being held there by capillarity. The pores of the membrane are assumed to be large compared to the dimensions of the reactants so that diffusion is not hindered. Membranes can be chosen for the reactor that are either hydrophilic (wet by the aqueous phase) or hydrophobic (wet by the organic phase). We consider here only the latter case.

The model we present here is analogous to that developed by Evans and Palmer (1981) to describe batch-phase–transfer reactions in which the interphase mass-transfer resistance plays a significant role. Their model was employed to predict conversion versus time for diaphragm cell experiments as a test of the phase-transfer mechanism (Evans, 1983).

A diagram of a differential length of the isothermal membrane reactor is shown with coordinate axes in Fig. 19. The origin of the axial coordinate is at the inlet of the reactor. The origin of the coordinate orthogonal to the direction of flow is at the aqueous/organic interface. Some of the organic phase reactant RX entering the organic flow channel with the feed reacts with reactant ion pair QY (active catalyst) that has diffused through the membrane from the aqueous/organic interface to form product RY and product ion pair QX (spent catalyst).

Also, some of the organic-phase reactant diffuses into the membrane to react with the counterdiffusing ion pair. Because the model must account for both these contributions to the net extent of conversion, both membrane and flow channel mass balances must be solved simultaneously.

Several assumptions were made to minimize complexity without sacrificing physical insight. Cocurrent plug flow in the channels adjacent to the membrane in the flat-sheet reactor was assumed; this assumption results in a relatively simple system of one-dimensional convection and reaction flow channel mass balances. However, flow channel mass-transfer resistances that contribute to the overall interphase resistance are not accounted for in the model. Second, ion-exchange equilibrium is assumed to exist at the oganic/aqueous interface, which resides on the aqueous side of the membrane. Finally, it is assumed here that the two ion pairs have equal diffusion coefficients and that the organic-phase reactant and product have equal diffusion coefficients.

Flow channel mass balances result in a system of seven first-order ODEs. Three are for the aqueous-phase ions (X^-, Y^-, and Q^+). Four are for the organic-phase species (the two ion pairs, QX and QY, and the organic-phase reactant and product, RX and RY). These mass balances with their requisite boundary conditions are given in dimensionless form by Eqs. (7) through (13):

$$0 = -\frac{dY^*}{d\hat{x}} + \frac{\Lambda}{FR}\frac{d\overline{QY}^*}{d\hat{z}}\bigg|_{\hat{z}=\theta} \qquad Y^*(0) = 1.0 \qquad (7)$$

$$0 = -\frac{dX^*}{d\hat{x}} + \frac{\Lambda}{FR}\frac{d\overline{QX}^*}{d\hat{z}}\bigg|_{\hat{z}=\theta} \qquad X^*(0) = 0 \qquad (8)$$

$$0 = -\frac{dQ^*}{d\hat{x}} + \frac{\Lambda}{FR}\left(\frac{d\overline{QY}^*}{d\hat{z}} + \frac{d\overline{QX}^*}{d\hat{z}}\right)\bigg|_{\hat{z}=\theta} \qquad Q^*(0) = \frac{ACR}{FR} \qquad (9)$$

$$0 = -\frac{dQY^*}{d\hat{x}} - \Lambda\frac{d\overline{QY}^*}{d\hat{z}}\bigg|_{\hat{z}=1} -R^* \qquad QY^*(0) = 0 \qquad (10)$$

$$0 = -\frac{dQX^*}{d\hat{x}} - \Lambda\frac{d\overline{QX}^*}{d\hat{z}}\bigg|_{\hat{z}=1} +R^* \qquad QX^*(0) = OCR \qquad (11)$$

$$0 = -\frac{dRX^*}{d\hat{x}} - \eta\Lambda\frac{d\overline{RX}^*}{d\hat{z}}\bigg|_{z=1} -R^* \qquad RX^*(0) = 1.0 \qquad (12)$$

$$0 = -\frac{dRY^*}{d\hat{x}} - \eta\Lambda\frac{d\overline{RY}^*}{d\hat{z}}\bigg|_{\hat{z}=1} +R^* \qquad RY^*(0) = 0 \qquad (13)$$

The dimensionless rate expression is given by

$$R^* = RX^*QY^* - \frac{1}{K^{eq}} RY^*QX^* \tag{14}$$

The dimensionless groups that arise from these equations are listed in Table 2. FR is the ratio of the molar feed rate of aqueous reactant to organic reactant. ACR is the ratio of the molar feed rate of catalyst introduced with the aqueous phase to the molar feed rate of organic-phase reactant. OCR is the ratio of the molar feed rate of catalyst introduced with the organic phase (in spent form) to the molar feed rate of organic-phase reactant. Λ is a dimensionless mass transfer rate that is best thought of as a measure of the relative rates of diffusion through the membrane to convection through the flow channels. η is the ratio of the diffusivities. ω is the space-time for the organic flow channel made dimensionless with a characteristic reaction time. Finally, K^{eq} is the equilibrium constant for the organic-phase displacement reaction.

The membrane mass balances result in a set of four second-order ODEs. These are given in dimensionless form in eqs. (15) through (18) with the necessary two boundary conditions per equation.

$$0 = D^{ip} \frac{d^2 \overline{QX}}{dz^2} + R \qquad \overline{QX}(0) = E^X(Q)(X) \qquad \overline{QX}(\delta) = QX \tag{15}$$

$$0 = D^{ip} \frac{d^2 \overline{QY}}{dz^2} - R \qquad \overline{QY}(0) = E^y(Q)(Y) \qquad \overline{QY}(\delta) = QY \tag{16}$$

$$0 = D^s \frac{d^2 \overline{RX}}{dz^2} - R \qquad \frac{d\overline{RX}}{dz}\bigg|_{z=\theta} = 0 \qquad \overline{RX}(\delta) = RX \tag{17}$$

$$0 = D^s \frac{d^2 \overline{RY}}{dz^2} + R \qquad \frac{d\overline{RY}}{dz}\bigg|_{z=\theta} = 0 \qquad \overline{RY}(\delta) = RY \tag{18}$$

The dimensionless rate term is given by

$$R^* = RX^*QY^* - \frac{1}{K^{eq}} RY^*QX^* \tag{19}$$

These equations contain terms accounting for diffusion and, because the membrane is wet by the organic phase, the displacement reaction. The boundary condition at $z = \delta$ imposes continuity of concentration. The boundary condition at $z = 0$ imposes no flux for the organic-phase reactant and product because

Table 2 Dimensionless Groups Appearing in
Phase-Transfer Catalysis Membrane Reactor
Model

$$FR = \frac{q^{aq}Y^0}{q^{org}RX^0}$$
  $$K^{eq} = \frac{k^f}{K^{rev}}$$

$$ACR = \frac{q^{aq}Q^0}{q^{org}RX^0}$$
  $$\omega = \frac{k^f RX^0 V^{org}}{q^{org}}$$

$$OCR = \frac{QX^0}{RX^0}$$
  $$\zeta = E^Y RX^0 \left(\frac{q^{aq}}{q^{org}}\right)^2$$

$$\Lambda = \frac{D^{ip} A_m}{\delta q^{org}}$$
  $$K^{sel} = \frac{E^Y}{E^X}$$

$$\eta = \frac{D^s}{D^{ip}}$$
  $$Da = \frac{k^f RX^0 \delta^2}{D^{ip}}$$

these species are not water soluble. Ion-exchange equilibrium is quantified in terms of the extraction relationships

$$[QY] = E^Y [Q] [Y] \qquad (20)$$

$$[QX] = E^X [Q] [X] \qquad (21)$$

where E^X and E^Y are the thermodynamic extraction constants for the two ion pairs.

Additional dimensionless groups associated with membrane-phase reaction and diffusion (see Table 2) include Da, a Damkohler number measuring the relative rates of reaction to diffusion in the membrane. ζ is a catalyst partitioning parameter that measures the extent to which the catalyst tends to partition into the organic over the aqueous phase. Finally, K^{sel} is the catalyst selectivity constant. This parameter measures the tendency of the catalyst to extract reactant anion over product anion.

A satisfactory solution to the problem was found by employing a regular perturbation technique (Da as the small parameter) to solve the membrane equations. This analytical solution was used to evaluate the flux expressions between the membrane and the flow channels. Finally, a Runge-Kutta routine was used to integrate the flow channel mass balances. Model predictions are compared with experiments in the next section.

B. Experimental Demonstration

We present here the results of experiments carried out to demonstrate reactor performance and to verify the validity of the assumptions made in developing the theoretical model. Bromooctane in the solvent chlorobenzene was reacted with aqueous iodide at 40°C to form the displacement products, iodooctane in chlorobenzene and aqueous bromide. The catalyst, tetrabutylammonium ion, was introduced as the bromide salt (spent form) in the organic feed. In all experiments, the organic feed solution used was 0.5 M in bromooctane, 0.05 M in TBABr, and 0.1 M in tetradecane (as a gas chromatographic standard) in the chlorobenzene. The aqueous solution was 2.0 M KI. Ion-pair concentrations were measured by potentiometric titration of the halide ions. Bromooctane and iodooctane concentrations were measured by gas chromatography.

To compare experiment with theory, it was necessary to evaluate under experimental conditions the dimensionless groups governing reactor performance. Therefore, independent, well-characterized, dispersed-phase experiments were carried out to determine the extraction constants of the two ion pairs in chlorobenzene, the forward displacement reaction rate constant, and the displacement reaction equilibrium constant. A list of these values is given in Table 3. Diaphragm cell experiments were run to characterize the membranes and to measure the diffusion coefficient of the TBABr in chlorobenzene. A literature value for the diffusion coefficient of bromooctane in chlorobenzene was used (Evans, 1983).

Experiments were run using the laboratory-scale flat-sheet reactor diagrammed in Fig. 20. The reactor was fashioned from blocks of aluminum into which inlet and outlet ports, flow distribution manifolds, and flow channels were machined. The flowing streams were isolated by bolting the halves of the reactor together with the membrane sandwiched between them. The membrane material selected for these studies was Goretex, a microporous film of polytetrafluoroethylene manufactured by W. L. Gore Associates.

Table 3 Values of Physical Parameters Used to Evaluate Dimensionless Groups

---

$E^I = 11.4 \ 1/M$

$E^B = 0.064 \ 1/M$

$k^f = 0.0010 \ 1/M \cdot s$

$K^{eq} = 0.24$

$\dfrac{D^{ip}}{\delta^{eff}} = 1.1 \times 10^{-4} \ cm/s$

$D^s = 1.5 \times 10^{-5} \ cm^2/s$ [a]

$D^{ip} = 0.58 \times 10^{-5} \ cm^2/s$ [a]

---

[a]From Evans (1983).

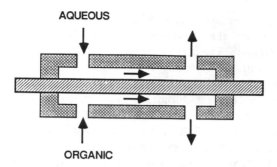

Figure 20 Schematic of a laboratory-scale demonstration membrane reactor. Flow channel dimensions: 0.05 cm deep; 20 cm long; 10 cm wide. Membrane material, Goretex (W. L. Gore, Assoc., Elkton, MD): nominal thickness, 0.025 cm; porosity, 61%; tortuosity, 1.7.

The reactant flows to the reactor were driven by constant-head tanks. The reactor was operated with the aqueous phase in the reactor at a slightly higher pressure (about 3 inches of water) than the organic phase. In this way, the non-wetting aqueous phase formed a barrier at its side of the membrane that prevented the organic phase from leaking from the membrane and being entrained in the aqueous stream. No entrainment of the organic phase in the aqueous stream effluent was detected in any of the experiments.

Steady-state reactor results are shown in Fig. 21 for two reactor runs, each 3 days in duration. The experimental variable was the organic flow rate. The values of two dimensionless groups, the space-time ω and the area A^*, depend on the organic flow and both are shown on the abscissa of Fig. 21. In all cases, the aqueous flow was maintained equal to the organic flow. In this way, the groups depending on the flow rate ratio were unchanged. Once operating conditions were established, the reactor was allowed to reach steady state, and several data points were then taken over a period of several hours to determine the conversion. Experimental results agree quite favorably with theoretical predictions, but in all cases experimental conversions are slightly below theoretical values. The most likely explanation is that resistance to mass transfer in the flow channels is not negligible; flow channel resistance is not accounted for in the model.

"Ideal reactor" performance is also plotted on Fig. 21. In the ideal reactor, ion-exchange equilibrium between the two flowing streams is imposed because the "ideal" membrane is assumed to offer no resistance to mass transfer. The difference between the predicted conversion in the real versus the ideal reactor represents the loss in reactor productivity suffered because of the mass-transfer resistance of the real membrane. This "mass-transfer penalty" is substantial for this reacting system in this flat-sheet reactor. However, the productivity of a membrane reactor relative to the ideal case can be greatly improved by employ-ing hollow-fiber membrane modules characterized by very high specific interfa-cial areas (Stanley et al., 1987).

Steady-state data were collected to verify the predicted dependence of the reactor performance on flow rate ratio; these results along with theoretical predictions are shown in Fig. 22. Again, the data fall a bit lower than predicted, but the dependence of conversion on flow rate ratio is confirmed. The decrease in conversion with increasing feed rate ratio FR (best thought of as increasing stoichiometric excess of the aqueous reactant) is a surprising result. The conver-sion falls because the distribution of catalyst is shifted out of the organic phase as the aqueous flow rate is increased.

In this section we have presented a new reactor concept for carrying out reac-tions between aqueous- and organic-phase reactants requiring the use of phase-transfer agents. Reactant streams are contacted by flowing along opposite sides

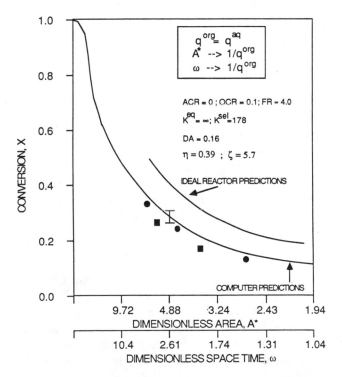

Figure 21 Steady-state performance of a membrane reactor versus flow rate of organic phase. At each steady-state point the aqueous flow equals the organic flow.

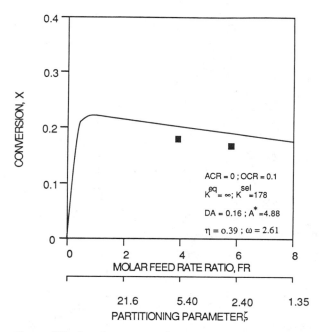

Figure 22 Steady-state performance of membrane reactor versus aqueous flow rate at fixed organic flow rate.

of a porous membrane that serves to stabilize the phase interface at a fixed position. Because intimate contacting is achieved without the formation of a dispersion, this device obviates the emulsification and coalescence problems frequently encountered in conventional dispersed-phase reactors. Also, because the interfacial area is a fixed, specified parameter, the membrane reactor can be operated with greater flexibility (i.e., over a greater range of flow rate or phase ratios).

V. CONCLUSIONS

Bioconversion via membrane reactors is a proven technology. All the steps involved in conventional bulk-phase processing, including liquid-liquid extraction, homogeneous catalysis, and phase contacting, can be adapted to two-dimensional membrane operations. In this chapter we have highlighted the considerable advantages inherent in membrane bioprocesses, with special emphasis on systems involving liquid-liquid or multiphase extraction operations and simultaneous catalytic conversion.

The reasons that membrane technology provides so many powerful possibilities for application in bioconversions result from the contacting properties of membranes per se:

1. Transport within the membrane structure can proceed by convection as well as diffusion, whereas nonmembrane porous supports are limited to diffusive transport alone.
2. The membrane structure allows immiscible liquids to be contacted with a fixed interface location, thus obviating some of the problems of conventional liquid-liquid contacting, namely emulsion formation and phase separation.
3. The interior of a membrane phase can be accessed through either bounding plane, that is, the two membrane interfaces provide an additional degree of freedom that allows integration of process steps in process integration.

It is this last feature that mimics the behavior of natural membranes, the ultimate in biochemical reactor design.

VI. NOMENCLATURE

A_m reactor membrane area
A^* dimensionless membrane area
ACR aqueous catalyst feed ratio
Da Damkohler number (dimensionless thickness)
D_{eff} effective diffusivity of reactant in the catalytic membrane
D^{ip} species i–ion-pair complex diffusivity
D^s diffusivity of organic reactant and product
D^{Sp} reactant–ion-pair complex effective diffusivity through the immobilized liquid membrane
D^{Pp} product–ion-pair complex effective diffusivity through the immobilized liquid membrane
E^s extraction constant for reactant
E^P extraction constant for reaction product
E^X extraction constant for ion X
E^Y extraction constant for ion Y
FR reactant feed rate ratio
k^f forward reaction rate constant for displacement reaction
k^r reverse reaction rate constant for displacement reaction
k_o first-order reaction rate constant
K^{eq} displacement reaction equilibrium constant
K^{sel} selectivity constant
l thickness of immobilized liquid membrane
L length of reactor
OCR organic catalyst feed ratio
q^{aq} aqueous flow rate
q^f flow rate of feed stream
q^{org} organic flow rate

q^P flow rate of the product stream
Q concentration of catalyst in aqueous phase
Q^O ion-pair-forming agent (carrier) concentration in the feed stream entering the reactor
Q^* dimensionless concentration of catalyst in aqueous phase
QX concentration of spent catalyst
QX^O feed concentration of catalyst in organic phase
QY concentration of active catalyst
QY^* dimensionless concentration of active catalyst
RX concentration of organic reactant
RX^O feed concentration of organic reactant
RX^* dimensionless concentration of organic reactant
RY concentration of organic product
RY^* dimensionless concentration of organic product
V^{org} volume of organic phase flow channel
x coordinate in axial direction
$\hat{x}$ dimensionless coordinate in axial direction
X concentration of aqueous product
X^* dimensionless concentration of aqueous product
Y concentration of aqueous reactant
Y^O feed concentration of aqueous reactant
Y^* dimensionless concentration of aqueous reactant
z coordinate normal to axial
$\hat{z}$ dimensionless coordinate normal to axial
δ membrane thickness
Λ dimensionless mass-transfer rate
η diffusivity ratio
ζ catalyst partitioning parameter
ω dimensionless space-time

REFERENCES

Axen, R., Porath, J., and Ernback, S., *Nature* 214:1302 (1967).
Balasingham, K., Warburton, D., Dunnill, P., and Lilly, M. D., *Biochim. Biophys. Acta* 276:250 (1972).
Boxer, G. E., and Everett, P. M., *Anal. Chem.* 21:670 (1949).
Brandstrom, A., *Advan. Phys. Org. Chem.* 15:267 (1977).
Carleysmith, S. W., and Lilly, M. D., *Biotechnol Bioeng.* 21:1057 (1979).
Carrington, T. R., *Proc. Roy. Soc., Ser. B* 179:312 (1971).
Chem. Week 37 (May 25) (1983).
Chiang, C., and Bennet, R. E., *J. Bacteriol.* 93:302 (1967).
Dehmlow, E. V., *Angew. Chem.* 16:493 (1977).
Dehmlow, E. V., and Dehmlow, S. S., *Phase-Transfer Catalysis.* Verlag Chemie, Weinheim (1980).

Evans, K. J., Interphase mass transport in phase-transfer catalyzed reactions, Ph.D. Thesis, University of Rochester (1983).

Evans, K. J., and Palmer, H. J., The importance of interphase transport resistances on phase-transfer catalyzed reactions, *AICHE Symp. Ser.*, 77(202) (1981).

Ford, W. T., and Tomoi, M., *Advan. Polym. Sci.* 55:49 (1984).

Freedman, A. H., *Pure Appl. Chem.* 58:857 (1986).

Gokel, G. W., and Weber, W. P., *J. Chem Educ.* 55:350 (1978).

Gordon, J. E., and Kutina, R. E., *J. Amer. Chem Soc.* 99:3903 (1977).

Herriott, A. W., and Picker, D., *J. Amer. Chem. Soc.* 97:2345 (1975).

Kiani, A., Bhave, R. R., and Sirkar, K. K., *J. Membr. Sci.* 20:125 (1984).

Kimura, S. G., Matson, S. L., and Ward, W. J., III, *Recent Developments in Separation Science*, Vol. 5. CRC Press, West Palm Beach, FL, p. 11 (1979).

Landini, D., Maia, A, Montanari, F., and Pirisi, F. M., *J. Chem Soc. Chem. Commun.* 950 (1975).

Landini, D., Maia, A., and Montanari, F., *J. Chem. Soc. Chem. Commun.* 112 (1977).

Landini, D., Maia, A., and Montanari F., *J. Amer. Chem. Soc.* 100:2796 (1978).

Lopez, J. L., Carrier -mediated transport in membrane reactors: Deacylation of benzylpenicillin, Ph.D. Dissertation, University of Pennsylvania (1983).

Matson, S. L., Membrane reactors, Ph.D. Dissertation, University of Pennsylvania (1979).

Matson, S. L., and Quinn, J. A., *Ann. N. Y. Acad. Sci.* 469:152 (1986).

Matson, S. L., Lopez, J., and Quinn, J. A., *Chem. Eng. Sci.* 38:503 (1983).

Melville, J. B., Mathematical modeling of solid-liquid phase-transfer catalysis, M.S. Thesis, University of Southern California (1983).

Melville, J. B., and Goddard, J. D., *Chem. Eng. Sci.* 40:2207 (1985).

Modin, R., *Acta Pharm. Suecica* 8:509 (1971).

Modin, R., and Schroder-Nielsen, M., *Acta Pharm. Suecica* 8:573 (1971).

Montanari, F., Landidi, D., and Rolla, F., *Topics Curr. Chem.* 101:147 (1982).

Park, J. M., Choi, C. Y., Seong, B. L., and Han, M. H., *Biotechnol. Bioeng.* 24:1623 (1982).

Qi, Z., and Cussler, E. L., *J. Membr. Sci.* 23:321 (1983a).

Qi, Z., and Cussler, E. L., *J. Membr. Sci.* 23:333 (1985b).

Reuben, R., and Sjoberg, K., *Chem. Tech.* (*Berlin*) 315 (May) (1981).

Rhodes, A., and Fletcher, D. L., *Principles of Industrial Microbiology*. Pergamon Press, Oxford (1966).

Schill, G., Isolation of drugs and related compounds by ion-pair extraction, in: *Ion Exchange and Solvent Extraction*, (J. A. Marinsky and Y. Marcus, eds.), Vol. 6. Marcel Dekker, New York (1974).

Schultz, J. S., *Recent Developments in Separation Science*, Vol. 3. CRC Press, Cleveland, OH, p. 243 (1977).

Smith, D. R., Lander, R. J., and Quinn, J. A., *Recent Developments in Separation Science*, Vol. 3. CRC Press, Cleveland, OH, p. 225 (1977).

Stanley, T. J., Advances in membrane reactors with applications to phase-transfer catalysis and dilution effects, Ph.D. Thesis, University of Pennsylvania (1986).

Stanley, T. J., and Quinn, J. A., Phase-transfer catalysis in a membrane reactor, *Chem. Eng. Sci.* 42:2313 (1987).

Stanley, T. J., Matson, S. L., and Quinn, J. A., Preparative organic synthesis via phase-transfer catalysis in a hollow fiber membrane reactor, to be submitted. (1990).

Starks, C. M., *J. Amer. Chem. Soc.* 93:195 (1971).

Starks, C. M., and C. Liotta, *Phase-Transfer Catalysis. Principles and Techniques.* Academic Press, New York (1978).

Starks, C. M.,and Owens, R. M., *J. Amer. Chem. Sco.* 95:3613 (1973).

Stinson, S. C., *C&E News* 27 (Feb. 17) (1986).

Vandamme, E. J., and Voets, J. P., *Advan. Appl. Microbiol.* 17:311 (1974).

Ward, W. J., III, *Recent Developments in Separation Science*, Vol. 1, CRC Press, Cleveland, OH, p. 153 (1972).

Way, J. D., Noble, R. D., Flynn, T. M., and Sloan, E. D., *J. Membr. Sci.* 12:239 (1982).

Weber, W. P., and Gokel, C. W., *Phase-Transfer Catalysis in ORganic Synthesis.* Springer-Verlag, New York (1977).

Yee, H. A., Palmer, H. J., and Chen, S. H., Surface deactivation and solbuilization in solid-liquid phase-transfer catalysis. Paper 49C AICHE National Meeting, Miami (1986).

4

Continuous Removal of Ethanol from Bioreactor by Pervaporation

Heinrich Strathmann

Universität Stuttgart
Stuttgart, Federal Republic of Germany

Wilhelm Gudernatsch

Fraunhofer-Institut für Grenzflächen- und Bioverfahrenstechnik
Stuttgart, Federal Republic of Germany

I. INTRODUCTION

The separation of products from other bioreactor constituents is often a difficult and costly step in large-scale industrial bioprocesses. Particularly when products or by-products have an inhibitory effect on the production rate, their continuous selective removal generally leads to significantly increased conversion rates and improved overall process economics. Recently there has been an increasing interest in the fermentation of organic solvents like ethanol or butanol from renewable resources, such as sugar, starch, or cellulose. These so-called biosolvents serve as fuel and chemical feedstock. At present, however, biosolvents are not competitive with petrochemical products because of high raw material and production costs. The price difference is expected to decrease over the next decade, however, because of possible improvements in the fermentation efficiency on one hand and increasing crude oil prices due to the disappearance of cheaply exploitable sources on the other hand.

Today, fermentation is mostly carried out as a batch process, which is effected by low productivity and high labor costs. This also holds true for the production of bioethanol, which is at present the most promising solvent usable as fuel or chemical feedstock. A substantial increase in bioconversion rates and thus in the overall production efficiency can be achieved by a continuous operation at higher cell densities. This can be achieved by a continuous feed of concentrated nutrients and a simultaneous removal of the product.

Several techniques based on semipermeable membranes are under consideration to achieve a continuous fermentation with high conversion rates. A conventional fermentor can be coupled to a membrane separation unit containing a micro- or ultrafiltration membrane as selective barrier. Microorganisms and other high-molecular-weight materials are retained when the fermentation broth is cycled through the membrane separator, as indicated in Fig. 1 (1). However, the product, that is, the ethanol, is obtained in relatively low concentration in a mixture with nutrients and other low-molecular-weight components in the permeate and requires further concentration and purification steps.

A selective removal of ethanol from the fermentor and its simultaneous concentration can be achieved by a membrane process referred to as pervaporation. In this chapter, the fundamentals of this process and its application in ethanol recovery using a composite hollow-fiber membrane are described and its economics illustrated in a brief cost analysis.

II. FUNDAMENTALS OF PERVAPORATION

Pervaporation is a membrane separation process that combines the separation of volatile components from a liquid mixture by evaporation and permeation

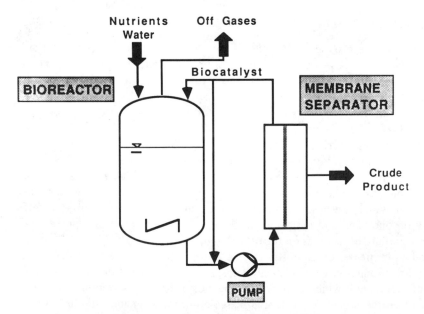

Figure 1 Product recovery by means of a membrane filtration unit.

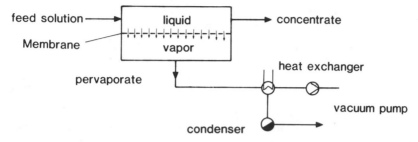

Figure 2 Principles of pervaporation.

through a semipermeable membrane. The principle of the process is illustrated in Fig. 2, which shows a liquid mixture separated from a gas phase by a selective membrane. The transmembrane flux of the various components is determined by their partial pressure gradient across the membrane and their permeability in the membrane matrix. Therefore, in pervaporation the flux of the various compounds in a mixture and thus their separation is determined not only by differences in their vapor pressure but also by the permeability of the membrane.

A. Fluxes and Driving Forces in Pervaporation

The mass transport in pervaporation can be broken down into three consecutive steps:

1. Sorption of components from a liquid phase at the membrane surface facing the feed solution
2. Diffusion of the sorbed components through the polymer matrix
3. Desorption and evaporation from the polymer matrix into the vapor phase on the permeate side of the membrane

The molar flux of a component $J_{n,i}$ in a homogeneous barrier, such as a pervaporation membrane, is given by (2)

$$J_{n,i} = L_i \, \mathrm{grad} \, \mu_i M \tag{1}$$

where $\mu_i M$ is the chemical potential of component i in the membrane and L_i is a phenomenological coefficient referring to the permeability of the component i in the polymer matrix.

The chemical potential is given by

$$\mu_i M = \mu_i M^0 + RT \ln a_i M \tag{2}$$

where $\mu_i M^0$ is the standard chemical potential, R the gas constant, and T the absolute temperature; $a_i M$ is the activity of component i in the membrane.

Considering only gradients perpendicular to the membrane surface, combination of eqs (1) and (2) leads to

$$J_{n,i} = L_i RT \; \frac{d \ln a_i M}{dx} \tag{3}$$

and

$$J_{n,i} = \frac{L_i RT}{a_i M} \; \frac{da_i M}{dx} \tag{4}$$

respectively, where x is the directional coordinate perpendicular to the membrane surface.

Assuming a linear relationship between membrane cross-sectional coordinate x and the activity a_i, integration of eq. (4) leads to

$$J_{n,i} = \frac{L_i RT}{\overline{a}_i M} \; \frac{\Delta a_i M}{\Delta x} \tag{5}$$

where $\overline{a}_1 M$ is an average activity of component i in the membrane, $\Delta a_i M$ is the activity difference of component i between the membrane feed and permeate side, and Δx is the thickness of the membrane.

Assuming local equilibrium between the membrane phase, the feed solution, and the permeate mixture, the activity of component i in the membrane can be related to its vapor pressure and its concentration, respectively, in the outer phases by

$$a_i = a_i M = \frac{P_i}{P_i 0} = f_i X_i = f_i M \; X_i M \tag{6}$$

where a_i and $a_i M$ are the activities, X_i and $X_i M$ the molar fractions, and f_i and $f_i M$ the activity coefficient of component i in the outer phase and the membrane, p_i is its partial pressure, and p_i^0 its saturation pressure.

Assuming the activity coefficient is constant across the membrane, combination of eqs (5) and (6) leads to

$$J_{n,i} = \frac{L_i RT \; \Delta X_i M}{\overline{X}_i M \; \Delta x} = \frac{L_i RT f_i \; Y X_i}{\overline{X}_i M f_i M \; \Delta x} = \frac{L_i RT \; \Delta P_i}{\overline{X}_i M p_i 0 f_i M \; \Delta x} \tag{7}$$

The phenomenological coefficient L_i and the activity coefficients f_i and $f_i M$ can be related to more commonly used diffusion and distribution coefficients as follows:

$$\frac{L_i RT}{\overline{X}_i} = -D_i M \tag{8}$$

$$\frac{f_i}{f_i M} = K_i^l \tag{9}$$

and

$$\frac{1}{P_i^0 f_i M} = K_i^g \tag{10}$$

Where D_i^M is the diffusion coefficient of component i in the membrane, and K_i^l and K_i^g are the distribution coefficient of component i between the feed solution and the permeate, respectively, and the membrane phase. Combination of Eqs. (7) through (10) leads to

$$J_{n,i} = -D_i^M K_i^l \frac{\Delta X_i}{\Delta x} = -D_i^M K_i^g \frac{\Delta p_i}{\Delta x} \tag{11}$$

Thus the molar flux of component i through a membrane is determined by its diffusion coefficient in the membrane and the distribution between the outer phases and the membrane.

The product of the diffusion coefficient and distribution coefficient is generally referred to as the permeability coefficient P_i:

$$-D_i^M K_i = P_i \tag{12}$$

Thus,

$$J_{n,i} = P_i^l \frac{\Delta X_i}{\Delta x} = P_i^g \frac{\Delta P_i}{\Delta x} \tag{13}$$

The permeation selectivity of a membrane for two different components i and j of a mixture is generally expressed by the ratio of their permeability coefficients:

$$S_{i,j} = \frac{P_i}{P_j} \tag{14}$$

where P_i and P_j are the permeabilities of the components i and j. In general, P_i and P_j are not constant but a function of the feed solution composition.

In general, diffusion coefficients and distribution coefficients and thus permeabilities of the various components in the membrane are not constant but a function of the composition of the outer phases.

B. Definitions of Separation and Enrichment Factors

The selectivity of a pervaporation membrane as defined by Eq. (14) is a useful parameter for characterizing and selecting the proper membrane for a given separation problem. For designing and adapting the actual process to a specific problem, the separation factor and enrichment factor are more appropriate.

The separation factor $a_{i,j}$ for a binary system is defined as

$$a_{i,j} = \frac{X_i'' X_j'}{X_i' X_j''} \qquad (15)$$

Where $a_{i,j}$ is the separation factor of the binary mixture containing components i and j, X_i and X_j are the molar fractions of the components, and the indices ($''$) and ($'$) refer to the permeate and the retentate, respectively.

The enrichment factor β_i is defined as

$$\beta_i = \frac{X_i''}{X_i'} \qquad (16)$$

where β_i is the enrichment factor and X_i'' and X_i' are the molar fractions of component i in the permeate and the retentate, respectively.

For the binary system, $a_{i,j}$ can be related to $S_{i,j}$ by combination of Eqs. (12) through (15) and introducing the proper relation between molar fractions X and the partial pressures p:

$$a_{i,j} = S_{i,j} \frac{X_j' (X_i' f_i P_i^0 - X_i'' p'')}{X_j' (X_i' f_j P_j^0 - X_j'' P'')} \qquad (17)$$

Where $a_{i,j}$ is the separation factor of the pervaporation process, $S_{i,j}$ is the selectivity of the membrane, X_i' and X_j' are the molar fractions and f_i and f_j the activity coefficients of components i and j in the feed solution, p_i^0 and p_j^0 are their saturation vapor pressures, S_i'' and X_j'' are the molar fracitons of components i and j in the permeate, and p'' is the gas pressure on the permeate side.

Equation (17) indicates that the separation factor $a_{i,j}$ in pervaporation depends not only on the selectivity of the membrane $S_{i,j}$ but is also determined by the activity coefficients f_i and f_j of components i and j in the feed solution, their saturation vapor pressures p_i^0 and p_j^0, and the total vapor pressure on the permeate side of the membrane p''.

Assuming vanishing low pressure on the permeate side of the membrane, that is, $p'' = 0$, Eq. (17) leads to

$$a_{i,j} = S_{i,j} \frac{f_i P_i^0}{f_j P_j^0} \qquad (18)$$

Where $S_{i,j}$ represents the separation due to the selectivity of the membrane and $f_i p_i^0 / f_j p_j^0$ represents the separation obtained owing to evaporation. Thus the pervaporation separation factor $a_{i,j}$ may either be larger than the separation factor obtained by distillation, when $S_{i,j} > 1$, or smaller, when $S_{i,j} < 1$.

The separation factor $a_{i,j}$ can also be related to the enrichment factor β_i by

$$a_{i,j} = \beta_i \, \frac{1 - X_j'}{1 - \beta_i X_i'} \tag{19}$$

or

$$\beta_i = \frac{a_{i,j}}{1 + (a_{i,j} - 1)X_i'} \tag{20}$$

Where $a_{i,j}$ is the separation factor for a binary system with components i and j, β_i is the enrichment factor of component i in the permeate, and X_i' and X_j' are the weight fractions of components i and j in the retentate. It should be noted that the separation factor $a_{i,j}$ and the enrichment factor β_i are defined as always >1.

III. PERVAPORATION MEMBRANE AND MODULE DEVELOPMENT

A. Preparation of the Pervaporation Membrane

The most important part of the pervaporation process is the actual membrane. This membrane should have the required separation properties, that is, a high permeation rate for ethanol, CO_2, and other volatile organic by-products. The water permeability should be low, and salts, substrates, and microorganisms should be completely rejected. Furthermore, for economic reasons, the ethanol transmembrane flux rate should be as high as possible. Since the flux rate is roughly inversely proportional to the membrane thickness, the actual selective barrier should therefore be as thin as possible. Membranes with extremely thin selective barriers with the necessary mechanical strength are obtained by producing composite structures. These membranes consist of a 0.1-5 μm thick homogeneous polymer film deposited on a microporous support structure, which provides mechanical strength but has virtually no effect on membrane selectivity or transmembrane flux rate. These are exclusively determined by the polymer film. The concept of a composite membrane is shown schematically in Fig. 3.

Various procedures for making composite membranes are described in the literature (4). For the preparation of ethanol-selective membranes very often poly(dimethylsiloxane) or similar rubbery polymers are used as the actual selective barrier and poly(sulfone) is used as the microporous support. The scanning electron micrograph of Fig. 4 shows the cross section of a composite membrane consisting of poly(dimethylsiloxane) selective barrier and microporous poly(sulfone) support structure. The selection of the barrier polymer as well as that of the support structure is of prime importance for the performance of the

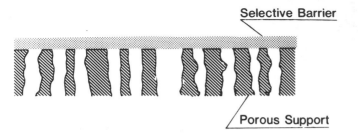

Figure 3 Concept of the composite membrane.

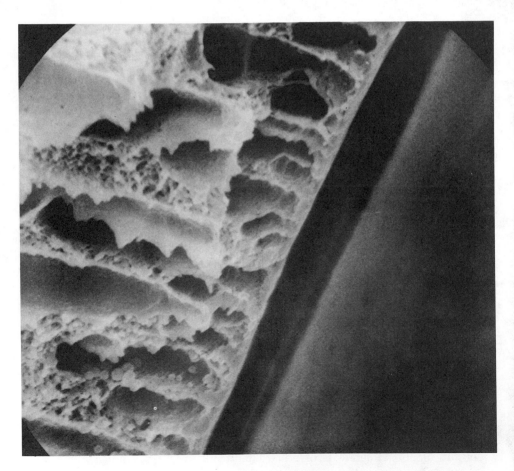

Figure 4 Scanning electron micrograph of an ethanol-selective composite membrane.

composite membrane. Several elastomers are described in the literature as having a good solvent selectivity over water (5). Most of the support structures are prepared from rigid glassy or partially crystalline polymers, which in general are water selective.

The selectivity of membranes with very thin selective barriers and low surface porosity of the support structure may be severely affected by the intrinsic properties of the support material. Thus the selectivity of the composite membrane is determined by

1. Selectivity of the selective barrier polymer
2. Porosity of the support structure
3. Selectivity of the support structure polymer

The flux rate through the membrane at given driving force is given by

1. Permeability of the selective barrier polymer
2. Thickness of the selective layer
3. Porosity of the support structure

Selection criteria for the barrier polymer as well as for the support structure polymer are described in the literature (5).

B. Membrane Module Design

To be useful in a large-scale technical separation process, any membrane must be incorporated into a suitable device, generally referred to as a module. In pervaporation, three module types are used today depending on the application. These are the plate-and-frame module, the spiral-wound modules, and the capillary membrane module.

The membrane is composed of a self-supported hollow-fiber or capillary microporous membrane with the actual selective barrier on the inside. A cross section of a capillary membrane is shown in the scanning electron micrograph of Fig. 5. The capillaries have an inner diameter of about 1–2 mm and are installed as a bundle into an outer shell tube, as indicated in Fig. 6. The feed solution is passed through the inside lumen of the capillaries, and permeate is collected in the outer shell tube.

The capillary membrane module provides some significant advantages. The flow geometry in the inside of a hollow fiber is well controlled, and concentration polarization effects as well as membrane fouling can be minimized by the proper feed flow velocity. Because of the absence of membrane support structure, the permeate pressure losses are relatively small; finally, the module provides a large surface area in a relatively small module volume, which leads to relatively low membrane production costs. Most of the experimental data reported in this chapter were obtained by using a capillary membrane module, with

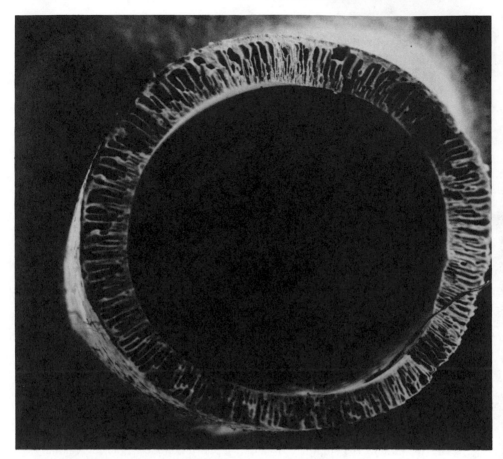

Figure 5 Cross section of a capillary composite membrane.

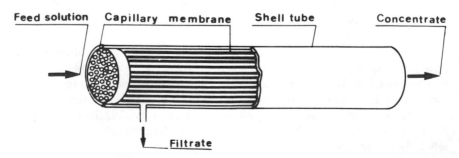

Figure 6 Capillary module.

the membrane made as a composite structure consisting of a porous poly(sulfone) substructure and a selective barrier about 1μm thick prepared from poly(dimethylsiloxane).

IV. PROCESS DESIGN

A. Basic Considerations

Solvents to be recovered from bioreactors usually have low concentrations. Their recovery therefore requires highly selective techniques to become economically feasible. However, the solvent-selective pervaporation membranes known at present show selectivities too low to be useful for concentrating the recovered product to usable levels.

Since pervaporation membranes can be made highly water selective, they can also be used for the depletion of an aqueous stream from its solvent load. A two-stage configuration could overcome the limitations of single solvent-selective membranes. Figure 7 shows this basic setup.

A solvent-selective pervaporation unit recovers the orgnic components from the feed cycle. The permeate from this stage is then stripped from its water content while flowing through a subsequent water-selective unit. This solvent recovery system acts like a virtual one-stage separator equipped with a much better membrane.

In the water-selective stage, membranes of different water selectivity characteristics can be employed in an advantageous way, optimizing the overall performance of the water-selective unit. The modules are operated in series. The

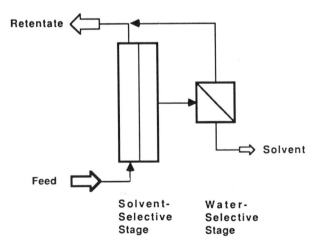

Figure 7 Two-stage pervaporation setup.

geometry and the flow velocity are adjusted in a way that each module is operated in its optimal selectivity range. Thus, the selectivity of an almost ideal membrane can be generated by arranging single selectivities in such a way that the enveloping curve describes the resulting overall selectivity. Figure 8 illustrates this method using three hypothetical membranes similar to those that already exist.

The diagram shows the selectivity characteristics of three different water-selective membranes M1, M2, and M3; w is the mass fraction of water in a binary ethanol-water mixture. The closer the characteristic of a single membrane comes to $w_{permeate} = 1.$ the more selective is the membrane. Experimental results show that different water-selective polymers have different feed concentration ranges (w_{feed}) of optimum selectivity. Therefore, water-selective membrane modules equipped with optimized membranes and operated in series as shown in the lower part of Fig. 8 can cover a much larger feed concentration range at optimum selectivity than modules equipped with just one type of membrane. The modules need only to be placed in such a way that the feed solution always meets membranes displaying optimum selectivity toward its momentary concentration. The effect of this configuration is the same as using one membrane with a selectivity characteristic of the enveloping curve.

B. Mass and Heat Transfer in Pervaporation

The driving force of pervaporation is induced by lowering the partial pressure of the permeands on the downstream side of the membrane. Thus, every permeand

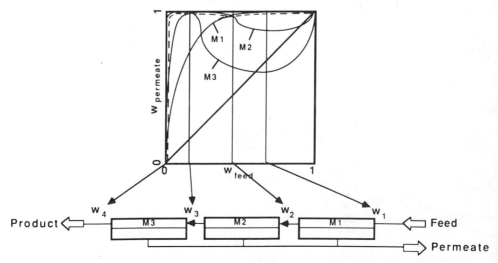

Figure 8 Effect of a module series equipped with different selectivity-optimized membranes.

must undergo a phase change. The required latent heat of evaporation is drawn from the feed solution. For the case of ethanol-water separation, the molar latent heats of evaporation are 201 and 539.1 kcal/kg, respectively (6). The evaporation of 1 kg/s of water therefore requires a power input of 2256 kW. This considerable amount of heat can be supplied as cheap waste heat from other processes.

The cooling energy for condensation, however, is an even more expensive input. For this reason, it should be carefully examined, whether a condensation is necessary. In the proposed configuration condensation of only the final product is needed. The first-stage permeate is sent into the second state, where it condenses on the water-selective membrane using only the heat of condensation of its water content to pervaporate it again. The permeate of the second state can be wasted as vapor since it consists of almost pure water. This design allows the direct reuse of the first stage's heat of evaporation in the second stage in the same way as in a rectification column.

Alternatively, the first-stage permeate vapors can be heated and stripped by water-selective vapor permeation. No major difference in permeation character-istics is expected between liquid and saturated vapor permeation (7).

In both cases, the minimum energy input is given by the heat of evaporation of the product and its final condensation.

C. Modes of Operation

The basic driving force of pervaporation is the permeand concentration gradient across the membrane. The feed side concentration depends on the partition of coefficient of the solvent, on its concentration, and on the temperature. The permeate side concentration must be significantly lower to allow a satisfactory flux. Maximum flux is obtained at a permeate pressure of zero (8).

For laboratory experiments, this condition can be set by applying a high vacuum to the permeate side of the module. This is usually done for such exper-imental purposes as characterizing the behavior of the system membrane-sol-vents-water.

However, from an economical point of view, absolute vacua for industrial processes moving high-vapor streams are too expensive to be taken into consid-eration. In reality, one is talking about technical vacua of about 20-30 mbar and more. The solvent recovery rate decreases remarkably when the permeate pres-sure is set to those values.

There are two more ways of lowering the permeate side concentration of the permeands: sweep gas pervaporation and thermopervaporation (9). The first works by sweeping a carrier gas through the permeate compartment, thus remov-ing the permeate vapors from the module. The other condenses the permeate vapors inside the module after a short distance at a cold wall.

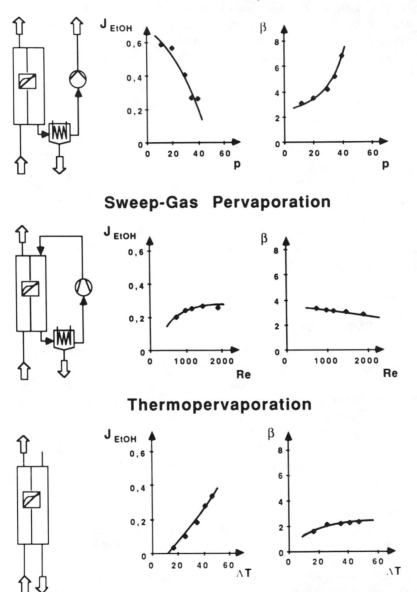

Figure 9 Comparison of the three basic operational modes of pervaporation.

To compare all three modes of operation, a hollow-fiber module was built equipped with a stainless steel housing for good heat conduction and permeate outlets on both ends. The feed was cycled through the hollow fibers. In vacuum pervaporation, the vacuum was applied on both outlets. In the sweep gas mode, one is used as gas inlet and the other as gas outlet. Finally, in thermopervaporation, the lower one was used as outlet and the upper one remained open to the ambient pressure. Figure 9 presents the results of experiments carried out using a synthetic mixture of 5 wt.% ethanol in water. The results obtained indicate that vacuum pervaporation is superior to sweep gas pervaporation and thermopervaporation in both flux and selectivity. Even at elevated permeate pressures of 30 mbar the vacuum flux is higher than the sweep gas flux. Thermopervaporation reaches a comparable flux at a temperature difference of 50°C, thus requiring a cooling temperature of -20°C for ethanol recovery purposes. However, sweep gas pervaporation could become an interesting alternative when used for pervaporative water stripping without subsequent condensation (10).

V. INTEGRATION OF PERVAPORATION INTO BIOCONVERSION PROCESSES

A. Description of the Separation Problem

Solvent fermentation broths consist of many chemically and physically different components. Their main solvent is water. Organic solvents, salts, and nutrients are physically dissolved. Gases are partially dissolved, partially forming bubbles. The microorganisms as well as different proteins are suspended in the solution.

Beside the product solvents, there are more organic trace components in the fermentation broth. The fermented solvents tend to inhibit the production activity of the microorganisms with increasing concentration. For reasons of production efficiency, they should be permanently removed from the fermentor, whereas nutrients and microorganisms should be kept inside. A further important point is the necessity to work under sterile process conditions with minimized thermal, chemical, and mechanical stress upon the microorganisms.

Figure 10 depicts the required permeation and rejection properties of the separation unit, using as an example the ethanol fermentation with *Saccharomyces cerevisiae*. It is evident that the separation unit must simultaneously carry out a complex set of single separation tasks to keep the fermentation process continuously in action. To operate under optimum conditions, the ethanol concentration in the fermentation broth should be adjusted as low as 5-8 wt.% relative to water (11,12). The ethanol concentration in the product, however, should be much higher to keep the further processing costs as low as possible. Therefore, a simultaneous preconcentration of ethanol integrated into the removal step is desirable.

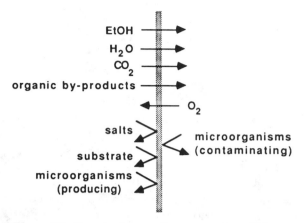

Figure 10 Required separation properties of the membrane.

B. Integration of the Pervaporation Process

Looking at possible separation processes suitable for the application described, pervaporation across highly permeable solvent-selective membranes seems to be the most attractive solution of the separation and concentration problem. It clearly avoids the high mechanical, thermal, or chemical stresses exerted upon the microorganisms by competitive processes, such as reverse osmosis, distillation, or solvent extraction, and holds the biggest potential of simultaneous preconcentration of the product (13). It is the only solvent-selective removal process able to keep the fermentation broth in the separator under conditions identical to those in the fermentor.

The process can be operated at low temperatures. Thus, the optimum fermentation temperature can be adjusted. Since the ethanol fermentation process is exothermic, it must be cooled (12). Pervaporation takes the heat of evaporation of the permeands out of the feed solution. It is therefore at least partially driven by the offal heat of the fermentation process. Furthermore, the gas exchange of the microorganisms can be carried out using the pervaporation module. The permeabilities of the solvent-selective membranes known at present toward CO_2 and O_2 are high. If proper pressure gradients for these two gases are established, they flow across the membrane.

Oxygenation of the fermentation broth is particularly useful during the start-up phase of a continuous fermentation. The cell density of yeasts grows under aerobic conditions. This can be used to breed the yeasts directly in the continuous fermentor up to the desired cell density. During production, a certain oxygen level increases the production rate (14). The removal of CO_2 across the membrane can be useful to control the foaming behavior of the fermentor.

In a mobile cell fermentation system, the fermentation broth in the separation unit is just as active as in the fermentor itself. Therefore, the inner space of the hollow fibers can be considered part of the fermentor vessel. Thus, the required fermentor volume can be reduced by the amount of the feedside volume of the pervaporation modules.

VI. SELECTED APPLICATIONS

The solvent selectivity of the selected membranes has been characterized using a synthetic feed mixture of ethanol and water. However, the central question of the membrane's reaction to contact with active fermentation broth had to be answered before any further evaluation and optimization could be reasonably done. It is well known from other membrane processes that fermentation broths tend to cause a particularly severe fouling of the membranes involved. On the other hand, poly(dimethylsiloxane) is a very biocompatible material. Furthermore, the surface of the active layer is very smooth (see Fig. 4). For these reasons, the interactions between polymer and fermentation broth could be expected to be small. To answer these questions, a laboratory-scale experiment using active fermentation broth was carried out.

A. Design of the Experimental Setup for Continuous Ethanol Fermentation

An integrated membrane bioreaction system with a fermentation volume of 350 ml was built. The hollow-fiber module had an area of 125 cm². Figure 11 shows the design of the process. The system designed as a closed loop consisting of the bioreactor, the membrane separation unit, and a circulation pump was set up. If run without bleed, the only input into the system is the nutrient solution and the only outflow is the solvent-enriched product vapor. Product recovery rates can be controlled by either the downstream pressure control valve V1 or the gas flow control valve V2. The permeated vapors are liquefied in the condenser. On-line measurement of the process parameters temperature (T), pH, oxygen concentration (O_2), solvent concentration (S), conductivity (L_f) and downstream pressure (p) provide sufficient information to control the process.

B. Ethanol Recovery

The following diagrams show characteristic results of vacuum pervaporation experiments obtained during 1 month of continuous fermentation of ethanol with *S. cerevisiae*. In Fig. 12, the partial flux of ethanol at a feed concentration of 5 wt.% is plotted versus the permeate pressure. The shaded rectangle shows an almost linear characteristic in the range above 20 mbar, that is, the range of economically feasible vacua. The ethanol recovery rate can be controlled by

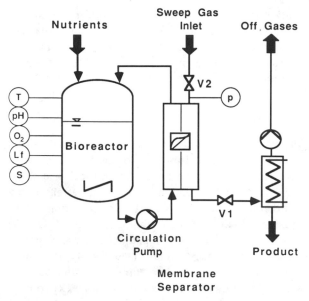

Figure 11 Experimental setup of a fermentation-pervaporation unit.

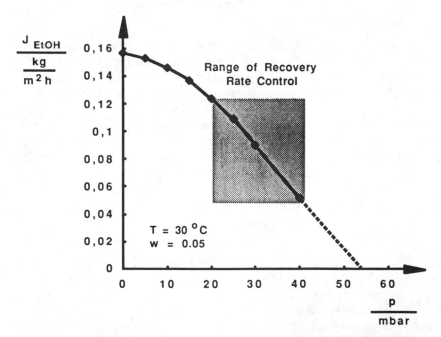

Figure 12 Dependence of the ethanol recovery rate on the downstream pressure.

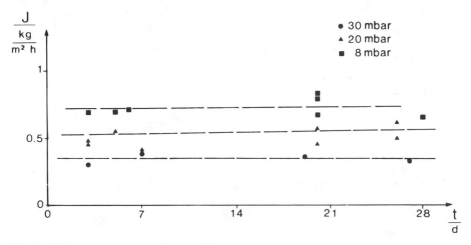

Figure 13 Long-term experiment: flux.

varying the downstream pressure in the given range. Figures 13 and 14 show the long-term behavior of the pervaporation experiments.

The average flux and selectivity at the feed concentration of 5 wt.% are slightly lower than the data obtained from synthetic ethanol-water mixtures but remain constant during the whole run. Table 1 gives the corresponding data. A slight fouling effect occurs immediately after contact of the membrane with the fermentation broth. No long-term fouling has been observed. From these

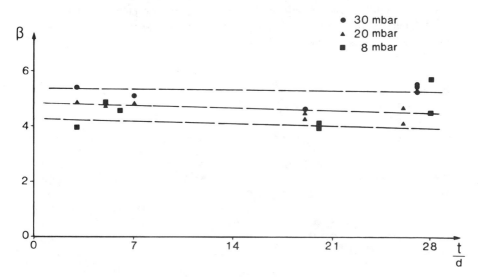

Figure 14 Long-term experiment: enrichment factor.

Table 1 Comparison of Pervaporation Using Synthetic Binary
Mixture and Active Fermentation Broth at 30°C and 10 mbar

Type of feed	Flux (kg/m^2·h)	β
Binary mixture	0.79	4.4
Fermentation broth	0.69	4.2

results it can be concluded that the separator used is able to accomplish the main
separation tasks, such as constant ethanol recovery, ethanol enrichment, and cell
recycling.

C. Copermeation of Organic By-productions

The microorganism *S. cerevisiae* converts glucose not only into ethanol and carbon dioxide but also into organic by-products, such as acetaldehyde, ethyl acetate, isobutanol, or acetic acid. These substances are much more dilute than
ethanol. In a conventional batch or flow-through fermentation system, their
concentrations are usually too low to seriously inhibit the activity of the microorganisms. In a recycling system, however, the danger of accumulation of those
trace components to inhibiting concentration levels exists. Therefore, the
copermeation behavior of some selected by-products has been investigated.
Table 2 gives the results of a representative experiment.

Table 1 proves that even at very low feed concentrations a copermeation of
trace components takes place. Methanol and ethyl acetate are concentrated from
a nondetectable level to a detectable one. Therefore, no enrichment factor can
be defined for those two components. Acetic acid, however, is not detectably
permeated. This is a drawback, because acetic acid has been identified as the

Table 2 Copermeation Behavior of Various Organic By-products

Component	w_f	w_p	β	J_{comp}(kg/m^2·h)
Ethanol	4.80E − 02	2.61E − 01	5.44	6.39E − 02
Acetaldehyde	2.63E − 04	2.51E − 03	9.52	6.14E − 04
Ethyl acetate	0.00E + 00	5.69E − 04	−	1.39E − 04
Methanol	0.00E + 00	2.36E − 04	−	5.78E − 05
Isobutanol	9.99E − 05	1.13E − 03	11.31	2.77E − 04
Methyl butanol	5.74E − 05	5.40E − 04	9.42	1.32E − 04
Acetic acid	7.32E − 05	0.00E + 00	0	0.00E + 00

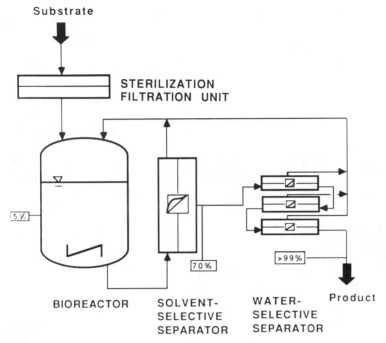

Figure 15 Membrane-controlled ethanol fermentation.

most inhibiting trace component when accumulated. If acetic acid cannot be removed simultaneously with ethanol from the fermentation broth, a feed-and-bleed operation must be installed to keep its concentration at tolerable levels (12).

VII. CONCLUSIONS AND FUTURE PERSPECTIVES

In addition to pervaporation there are other membrane processes, such as micro-filtration or membrane distillation, that can be utilized in a continuous fermentation of ethanol (13). However, pervaporation seems to be the most promising process in terms of overall economics. Microfiltration is limited to the separation of the biocatalyst only and requires such follow-up procedures as conventional distillation to achieve a truly continuous production process. Membrane distillation has technical problems, and its selectivity will always be limited by the vapor-liquid equilibrium. At least in this respect no further improvement is possible. Today, the selectivity obtained by pervaporation is only insignificantly better than that obtained by membrane distillation. A significant improvement in membrane performance is quite likely to be seen in the near future. Pervaporation membranes with transmembrane fluxes of more than 2–3 $kg/m^2 \cdot h$ and selectivities in excess of S = 50 under reasonable operating conditions have been developed on a laboratory scale (15) and should soon be commercially available. With this type of membrane, ethanol continuously removed from the fermentation broth can simultaneously be concentrated to about 70–80 wt.%. This ethanol-water mixture could then be completely dehydrated by pervaporation using a water-selective membrane, as indicated in Fig. 15, which shows a schematic diagram of a continuous ethanol fermentation process completely controlled by membrane processes, indicating three stages. In the first stage such substrates as glucose or lactose are sterilized by microfiltration; in the second stage ethanol is continuously removed from the fermentors by pervaporation through an ethanol-selective membrane, keeping the fermentor ethanol concentration below the process-inhibiting concentration of 5 wt.%. Finally, the ethanol is concentrated by pervaporation through a water-selective membrane to >99 wt.%.

REFERENCES

1. Flaschel, E., Wandrey, C., and Kula, M-R., Ultrafiltration for the separation of biocatalysts, *Advan Biochem Eng. Biotechnol* 26:73–142 (1983).
2. Strathmann, H., *Trennung von molekularen Mischungen mit Hilfe Synthetischer Membranen.* Steinkopff Verlag, Darmstadt, p. 15–35 (1979).
3. Néel J., Aptel, P., and Clement, R., Basic aspects of pervaporation, *Desalination* 53:297–326 (1985).
4. Cadotte, J. E., Evolution of composite reverse osmosis membranes, in: *Material Science of Synthetic Membranes* (D. R. Lloyd, ed). ACS Symposium Series 269, pp. 273–294 (1985).
5. Baker, R. W., and Blume, I., Permselective membranes to separate gases. *Chem. Tech. (Berlin)* (April) (1986).
6. Dubbel, *Taschenbuch für den Maschinenbau*, Band I. Springer Verlag, Berlin (1974).
7. Böddeker, K. W., Pervaporation durch Membranen und ihre Anwendung zur Trennung von Flüssigkeitsgemischen. Fortschrittsberichte VDI, Reihe 3: Verfahrenstechnik 129 (1986).
8. Paul, D., and Paciotti, J. D., Driving force for hydraulic and pervaporative transport in homogeneous membranes. *J. Polym. Sci., Polym. Phys. Ed.* 13: 1201–1214 (1975).
9. Aptel, P., and Néel, J. Pervaporation, in: *Synthetic Membranes: Science, Engineering and Applications.* NATO ASI–Series C, Mathematical and Physical Sciences, Vol. 181 (1986).
10. Cabasso, I., One stage separation by pervaporation. Paper presented at First Annual National Meeting of the North American Membrane Society, Cincinnati, OH (1987).

11. Nagashima, M., Azuma, M., Naguchi, S., Inuzuka, K., and Samejima, H., Continuous ethanol fermentation using immobilized yeast cells, *Biotechnol. Bioeng.* 26:992-997 (1984).
12. Maiorella, B., Blanch, M., and Wilke, C., Economic evaluation of alternative ethanol fermentation processes, *Biotechnol. Bioeng.* 24:1003-1025 (1984).
13. Mulder, M. H. V., and Smolders, C. A., Continuous ethanol production controlled by membrane processes, *Process Biochem.* (April) (1986).
14. Ry, D. D. V., Kim, Y. J., and Kim, J. H., Effect of air supplement on the performance of continuous ethanol fermentation, *Biotechnol. Bioeng.* 29: 12-16 (1984).
15. Nagase, Y., Mari, S., Matsui, K., and Uchikara, M., Synthesis of substituted polyacetylene/polydimethylsiloxane graft copolymers and the application for gas separation and pervaporation. Paper presented at the International Congress on Membranes and Membrane Processes, Tokyo, Japan (1987).

5
Perstraction

Masatoshi Matsumura

Institute of Applied Biochemistry, University of Tsukuba
Tsukuba, Ibaraki, Japan

I. INTRODUCTION

A number of fermentations exhibit product inhibition. This undesirable phenomenon limits the use of a concentrated substrate and results in product recovery becoming more difficult and more expensive. If high sugar concentrates, such as full-strength molasses, were applicable in a continuous fermentation, it would be possible to reduce not only the size of fermentation plants to obtain a desired output but also the cost of wastewater treatment.

Fermentations of alcohols such as ethanol and butanol are typical, showing serious product inhibition. Biomass alcohols are attracting a great deal of attention as a promising liquid fuel and chemical feedstock. However, practical application is prevented by their low productivities due to product inhibition and high energy expenditure for purification. Approaches to overcoming product inhibition are generally based on continuously removing the inhibiting compounds during fermentation. Consequently, the viability of biomass alcohols becomes dependent on the development of a low-cost, energy-efficient separation process for alcohols.

Many proposals have been made for separating alcohols from fermentation broth: stripping under reduced pressure (1,2), selective adsorption on solid adsorbents (3-6), solvent extraction (7-10), membrane separation based on reverse osmosis (11-14), membrane separation based on pervaporation (15-19), and so on.

We have been studying a separation process for alcohols based on a combination of membrane permeation and solvent extraction (20-22). This process is called perstraction. Dialysis may be considered a kind of perstraction using water as the extractant. In our early studies of the separation of ethanol, we used a hydrophilic hollow-fiber membrane and demonstrated its excellent extraction capacity in actual ethanol fermentation. Recently we developed another mode for perstraction. It was a separation method using a liquid membrane supported with a hydrophobic porous membrane. This new mode was applied to the separation of n-butanol, ethanol, isopropanol, and acetone and showed itself to have excellent performance, especially for n-butanol. This chapter explains the characteristics of perstraction, focusing on our experimental results.

II. PERSTRACTION USING A HYDROPHILIC HOLLOW-FIBER MEMBRANE

In previous extractive fermentations (7-10), solvent was supplied to the fermentor and came into direct contact with the fermentation broth. As a result, only nontoxic and nonemulsion-forming solvents could be used. This means that the selection of solvents becomes restricted. Furthermore, Crabbe et al. (23) showed that in a direct contact system yeast cells were attracted to liquid/liquid interfaces and this severely reduced the rate of ethanol extraction.

A promising method to overcome these problems in the extractive fermentation was the application of a membrane. When solvent extraction is carried out through a membrane, the accumulation of microorganisms on the solvent surface is prevented and any toxic effect of solvent is reduced. Many kinds of hollow-fiber modules have been developed for medical treatment. Using these hollow-fiber modules, it is possible to construct a recycling system that has several potential advantages, such as effective utilization of process water and nutrients (24). Even in such a recycling system, however, solvent accumulates in the medium at concentrations up to saturation. Therefore it has become necessary to develop a method of protection against solvent toxicity to maintain a free hand in the selection of solvents.

A. Selection of Solvent

1. Ethanol-Extracting Capacity

Using 40 organic solvents, Roddy (25) measured distribution coefficients for ethanol and water in a ternary system of solvent, water, and ethanol and obtained the separation factors of the solvents from these distribution coefficients. His results showed the order of extraction capacity of ethanol to be hydrocarbon $<$ ether $<$ ketone $<$ amine $<$ ester $<$ alcohol.

Solvent selection should be done based mainly on the distribution coefficient and separation factor, because these values strongly affect the size of extractor and the solvent requirement. At the same time, attention should also be paid to other physical properties that affect the costs of an entire process. For example, a solvent with low solubility in water should be employed to avoid an additional process for solvent recovery.

We regarded perstraction as a pretreatment process for distillation and looked for a suitable extractant among 25 organic solvents with high boiling points (20). The typical examples of distribution curve of ethanol between solvent and water are shown in Fig. 1. The ethanol distribution coefficients m, defined as the ratio of the concentration of ethanol in the solvent phase to that in the aqueous phase, were not constant, and they had a tendency to increase with increasing ethanol concentration in the aqueous phase. However, in the low ethanol concentration region up to 100 g/L, which is important for the extraction of a fermentation broth, the distribution coefficient can be considered constant. The distribution coefficients obtained in this concentration are also shown in Fig. 1. No solvent with a distribution coefficient of more than unity was found among the solvents tested. Honda et al. (9) showed that phenol derivatives, such as o-*tert*-butylphenol and o-isopropyl-phenol, had a high distribution coefficient, around 1.4. These solvents are well known for their high toxicity. Since much attention should also be paid to occupational health and safety, it would have been difficult to use them on a large scale.

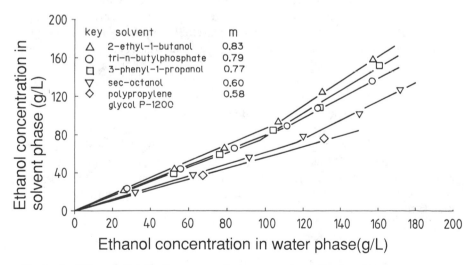

Figure 1 Ethanol distribution curves between water and solvents.

2. Solvent Toxicity to Ethanol-Producing Microorganisms

As mentioned previously, even in a system using a membrane, fermentation broth would contain the solvent at concentrations up to the saturation point. Therefore it was necessary to investigate the solvent toxicity to ethanol-producing microorganisms. The effect on cell growth was examined for 10 solvents with fairly high extracting capacity (20). The ethanol-producing microorganisms used in this experiment are well known to exhibit such characteristics as flocculation (*Schizosaccharomyces pombe* and *Saccharomyces uvarum*) (26), ethanol tolerance (*Saccharomyces cerevisiae sake*) (27), thermotolerance (*Candida brassicae*) (28), and high ethanol productivity (*Zymomonas mobilis*) (29). Table 1 shows the effect of the solvents on the growth of these typical ethanol-producing cells. Three solvents proved nontoxic to cell growth. Polypropylene glycol P-1200 was fairly viscous and formed an emulsion with aqueous ethanol solutions at concentrations below 100 g/L. 2-Ethyl-1,3-hexanediol was even more viscous than polypropylene glycol P-1200 and had a similar emulsifying tendency. Methyl crotonate exhibited desirable characteristics, such as low toxicity, fairly high extraction capacity, and nonemulsification, but its price, unfortunately, is too high to allow employment on an industrial scale. Other solvents, such as 2-ethyl-1-butanol, tri-*n*-butylphosphate, and the octanols are already employed as industrial solvents and have several superior characteristics for ethanol extraction. However, they are toxic to the microorganisms tested, especially 2-ethyl-1-hexanol and *n*- and *sec*-octanol, which completely inhibited cell growth even at a concentration of 0.05 vol.%.

Similar experiments on solvent toxicity were carried out by Playne et al. (31) and Honda et al. (9). Using a mixed culture of facultative anaerobic, acid-producing bacteria, Playne et al. investigated the toxicity of 30 organic solvents and found that 13 solvents were nontoxic. They were paraffins (C_6–C_{12}), phthalates, organophosphorous compounds, Freon 113, Aliquat 336, diiosamyl ether, and trioctylamine. They also found that the solvents were generally not toxic unless present at levels in excess of that which would be required to saturate the aqueous phase. Honda et al. investigated the toxicity of 12 solvents by using both *S. cerevisiae* and a strain PN13 obtained by protoplast fusion between *S. cerevisiae* and *Kluyveromyces lactis*. The aliphatic alcohols of carbon number above 10, such as 1-decanol, oleyl alcohol, and Guerbet alcohol (branched-chain alcohol of carbon 20), were found not to inhibit microbial activities, but the phenol derivatives (*o-tert*-butylphenol and *o*-isopropylphenol) showed drastic toxicities.

B. Protection Against Solvent Toxicity

Almost all solvents suitable for the extraction of ethanol are toxic to microorganisms. This means that the development of some sort of protection method

Table 1 Solvent Toxicity to the Growth of Typical Ethanol-Producing Cells[a]

Solvent	*Schizosaccharomyces pombe* IFO 0344	*Saccharomyces cerevisiae* IFO 0309	*Candida brassicae* IFO 1664	*Saccharomyces uvarum* ATCC 26602	*Zymomonas mobilis* IFO 13766
Methyl crotonate	++	++	++	++	++
Tri-*n*-butylphosphate	-	-	+	-	+
2-Ethyl-l-butanol	+	+	+	+	+
n-Octanol	-	-	-	-	-
Sec-octanol	-	-	-	-	-
2-Ethyl-l-hexanol	+	+	+	-	-
3-Phenyl-l-propanol	+	+	+	+	+
2-Ethyl-l,3-hexanediol	++	++	++	++	++
Isoeugenol	-	-	+	-	-
Polypropylene glycol P-1200	++	++	++	++	++

[a]-, +, and ++ indicate 0–19, 20–79, and 80–100% of the growth compared to solvent-free control.

against solvent toxicity would be essential to make up an efficient extractive fermentation system. We developed the following method (20).

An adsorbent with a strong affinity for a solvent was immobilized together with cells using Na alginate. The adsorbent trapped the toxic solvent molecules coming into the gel beads and reduced the solvent concentration in the gel beads to much less than that in the fermentation broth. Figure 2 shows the protection effect of Porapack Q (100–120 mesh, gas chromatographic support) against the toxicity of *sec*-octanol. Using fresh gel beads immobilizing both small numbers of yeast cells and Porapack Q, batch fermentation was carried out in a medium containing 0.1 vol.% *sec*-octanol, and changes in glucose and ethanol concentrations were monitored. The results obtained with adsorbent-free gel beads in solvent-free medium and with adsorbent-free gel beads in medium containing solvent are also shown in Fig. 2. This figure indicates that cell metabolism was totally inhibited in the adsorbent-free gel beads by *sec*-octanol at 0.1 vol.%, although in the gel beads containing Porapack Q the cell growth accompanying glucose consumption and ethanol accumulation began after a short time lag. This lag time may have been required to reduce the concentration of *sec*-octanol in the gel beads to a certain safety level for cell growth. The yield of ethanol was almost the same as that obtained with the adsorbent-free gel beads. This result means that the loss of ethanol by adsorption on Porapack Q was not serious.

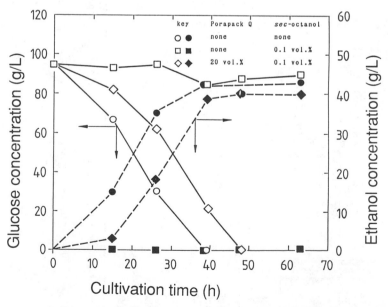

Figure 2 Protection effect of Porapack Q against the toxicity of *sec*-octanol.

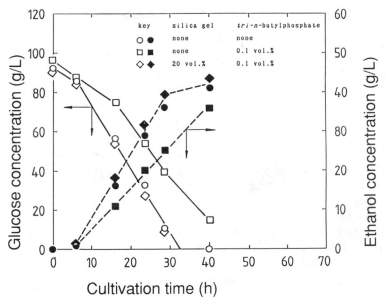

Figure 3 Protection effect of silanized silica gel against the toxicity of tri-*n*-butylphosphate.

Figure 3 shows the protective effect of silanized silica gel (70–230 mesh, liquid chromatographic support) against the toxicity of tri-*n*-butylphosphate. The experiment was carried out in the same way as in the case of *sec*-octanol. The toxicity of tri-*n*-butylphosphate was removed so quickly that there was no detectable lag in ethanol production or glucose consumption. These results were identical to those obtained in the solvent-free medium. It can be seen from this figure that gel beads alone were fairly effective in preventing the toxic effect of tri-*n*-butylphosphate.

Our protection method was also applied to extractive fermentation using *o-tert*-butylphenol and *o*-isopropylphenol as extractants (9). In this case a castor oil of 20 vol.% was employed as an absorbent of the solvents. The toxicity of *o-tert*-butylphenol was almost completely removed, but in the case of *o*-isoropyl-phenol the protective effect was not great enough.

The protector in gel beads is effective until it is saturated with solvent. Durability of the protective effect depends on the amount of protector in gel beads and on the concentration of solvent in the fermentation broth. Therefore, the extraction of ethanol through a membrane by using a solvent with low solubility in water is effective in prolonging the durability of the protective effect.

C. Ethanol Perstraction in a Hollow-Fiber Module

Hollow-fiber membranes have been made of various kinds of materials. Using hollow-fiber membranes made of cupro-ammonium cellulose, ethylenevinyl alcohol, polyvinyl alcohol, cellulose diacetate, polyacrylonitrile, and polysulfone, the tolerance of membranes to the solvents in Table 2 was checked in a preliminary experiment. Cuproammonium cellulose and polyvinyl alcohol were found to be fairly stable to these solvents. A hollow-fiber membrane made of cuproammonium cellulose (Cuprophan, Enka A.G., West Germany) was employed in our experiments on ethanol perstraction (21); the details of the hollow-fiber modules (Terumo Co. Ltd., Japan) are shown in Table 3. They are the kind used for blood treatment in artificial kidneys.

Figure 4 shows a schematic diagram of the experimental setup used to investigate the mass-transfer characteristics of ethanol in the module. Three solvents with fairly high ethanol distribution coefficients were employed, with water also used as a standard extractant. Ethanol-free extractant preheated to 30°C was fed into the inside of hollow fibers at a given feed rate and allowed to flow out through an overflow pipe. An aqueous ethanol solution of around 80 g/L at 30°C was fed along the outside of the hollow fibers. The two liquids contacted countercurrently through the membrane. Needle valves in the liquid flow lines controlled the pressure of both liquids. The inlet and outlet ethanol concentratins were measured in a steady state.

The following material balance equation holds under the conditions in which the solvent and aqueous ethanol solution flows are pistonlike in opposite directions, and the mass transfer of ethanol occurs only by diffusion through the membrane:

$$-d(Sx) = -d(Wy) = K_La(mC_{E/W} - C_{E/S}) \tag{1}$$

where S, x, and $C_{E/S}$ indicate the solvent molar flow rate per cross-sectional area, ethanol molar fraction, and ethanol molar concentration in solvent, respectively; W, y, and $C_{E/W}$ are for the aqueous ethanol solution. K_La is the overall volumetric mass-transfer coefficient, and Z is the length of the hollow fibers. The molar flow rates S and W can be treated as constant under conditions in which neither solvent nor water dissolves in each other, and the ethanol content in the feed aqueous ethanol solution is small. If these conditions are satisfied, Eq. (1) can be simplified to Eq. (2):

$$-Q_s \frac{dC_{E/S}}{dz} = -Q_w \frac{dC_{E/W}}{dz} = K_La(mC_{E/W} - C_{E/S}) \tag{2}$$

where Q_s and Q_w are the superficial velocities of solvent and aqueous ethanol solution, respectively. Using the following boundary conditions, the relative

Table 2 Physical Properties of Solvents Used as Extractant in Perstraction Experiments

Solvent	Distribution coefficient of ethanol, m (–)	Solubility of water in solvent, Sw/s (g/L)	Solubility of solvent in water, Ss/w (g/L)	Viscosity of solvent, μ (mPa·s)	Diffusion efficient of ethanol in solvent, D (m²/h)
2-Ethyl-l-butanol	0.83	44.0	8.50	4.91	1.26×10^{-6}
Tri-n-butylphosphate	0.79	64.5	0.37	2.95	3.38×10^{-6}
Sec-octanol	0.60	36.2	0.66	5.14	1.37×10^{-6}
Water	1.00	–	–	0.80	51.5×10^{-6}

Table 3 Specification of the Hollow-Fiber Modules Used in Perstraction

	Module type	
Properties	Terumo TE10	Terumo TH10
Contact area, m^2	1	1
Module diameter, mm	38	38
Module length, mm	170	230
Fiber diameter, μm	200	300
Fiber thickness, μm	11	16
Fiber number	9800	4900
Fiber material	Cuprophan	Cuprophan
Housing material	Polycarbonate	Polycarbonate

ethanol concentration in aqueous ethanol solution at the inlet $(C_{E/W})_1$ and at the outlet of the module $(C_{E/W})_2$ are given by Eq. (3):

$$Z = 0: \ C_{E/W} = (C_{E/W})_1 \qquad C_{E/S} = (C_{E/S})_1$$
$$Z = Z: \ C_{E/W} = (C_{E/W})_2 \qquad C_{E/S} = 0$$

$$\frac{(C_{E/W})_1}{(C_{E/W})_2} = \frac{\Psi e^{-\Psi(mK_LaZ/Q_W)}}{1 - [(1/m)(Q_W/Q_S)]e^{-\Psi(mK_LaZ/Q_W)}} \tag{3}$$

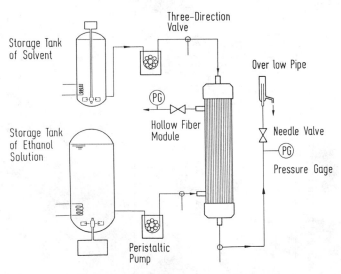

Figure 4 Experimental setup for ethanol perstraction.

where $\Psi \equiv 1 - (1/m)\ Qw/Qs$. As these assumptions were for the most part satisfied in the case of the three solvents shown in Table 2, the overall volumetric mass-transfer coefficient K_La in perstraction with these solvents was obtained from Eq. (3). When water is used as an extractant, a transfer of extractant may also occur. In such a case, K_La should be obtained from Lane's equation (32), in which the mass transfer of extractant was also taken into account.

Figures 5 and 6 show the effects of the inside and outside flow rates of hollow fiber Qs and Qw on the overall volumetric mass-transfer coefficient K_La, respectively. In the experiment on the effect of the inside flow rate Qs, water and *sec*-octanol were used as extractants, and the outside flow rate Qw was constant at 1.17 m/h. The dependence of K_La on Q_s differed with the kind of extractant. To explain this, we applied Leveque's equation (33), which was derived for the mass transfer from tube walls to laminar fluids.

The overall mass-transfer resistance is divided into the outside liquid film resistance of hollow fibers, the membrane resistnace, and the inside liquid film resistance. When the thickness of the liquid films and hollow-fiber membrane is very slight, these contact areas can be considered the same and the following relation holds:

$$\frac{1}{K_L} = \frac{1}{k_L} + \frac{1}{k_m} + \frac{1}{k_s} \tag{4}$$

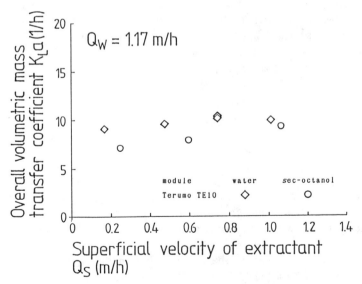

Figure 5 Effect of inside flow rate of hollow fiber on overall volumetric mass-transfer coefficient.

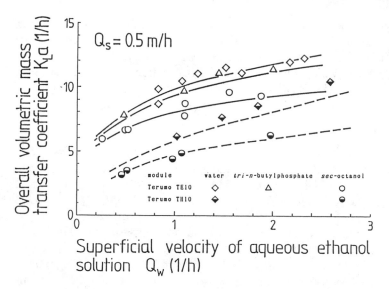

Figure 6 Effect of outside flow rate of hollow fiber on overall volumetric mass-transfer coefficient.

where K_L is the overall mass-transfer coefficient, and k_L, k_m, and k_s are the mass-transfer coefficients of the outside liquid film, the membrane, and the inside liquid film, respectively. Under the condition of constant outside flow rate Qw the values of $1/k_L$ and $1/k_m$ are independent of the inside flow rate Qs, and $1/K_L$ is proportional to $1/k_s$. If Leveque's equation given by Eq. (5) is applicable, the mass-transfer coefficient of the inside liquid film can be expressed by Eq. (6).

$$Sh = 1.61 \left(ReSc \, \frac{d}{Z} \right)^{1/3} \tag{5}$$

$$\frac{1}{k_s} = \frac{d^{1/3} Z^{1/3}}{1.61 D^{2/3}} \left(\frac{Q_s}{\epsilon} \right)^{-1/3} \tag{6}$$

where Sh and Re are the Sherwood number and the Reynolds number, D is the diffusion coefficient of ethanol in extractant, d the inner diameter of hollow fibers, and ϵ the ratio of the cross-sectional area of hollow fibers to that of the module. Therefore, $1/K_L$ is proportional to $Q_s^{-1/3}$.

Overall mass transfer coefficients of the Terumo TE10 module were obtained by dividing K_La in Fig. 5 by a (5190 L/m for TE10), and their reciprocals were plotted against $Q_s^{-1/3}$. As shown in Fig. 7, $1/K_L$ was almost proportional to

$Qs^{-1/3}$ and the slopes of these straight lines agreed fairly well with the calculated values from Eq. (6): water = 43 and *sec*-octanol = 110. These results showed that Leveque's equation was applicable to hollow fibers and that the difference in dependence of K_{La} on Qs indicated in Fig. 5 was due to the difference in diffusion coefficients of ethanol in the extractants (see Table 2).

Figure 6 shows the effect of the outside flow rate of hollow fiber Qw on K_{La} of the Terumo TE10 and TH10 modules. In this experiment the inside flow rate Qs was fixed at 0.5 m/h. A viscous extractant, such as *sec*-octanol, gave smaller K_{La} values than that of water, and the difference in K_{La} became greater with increasing Qw. Since in the low Qw region the resistance in the outside liquid film should be dominant, there was not so great a difference in K_{La}. In the high Qw region, the resistance in the outside liquid film decreased with increasing Qw, and the effect of the resistance in the inside liquid film appeared in K_{La}. Using these measured K_{La} values and Leveque's equation, we determined the mass-transfer coefficients of the outside liquid film and membrane. Wilson's plot [see Bohrer (34)] for Terumo TE10 and TH10 modules are shown in Fig. 8. The values of ($1/K_L$ - $1/ks$) were obtained for each of the measured K_{La} values, and an exponent of $Q w/(1 - \epsilon)$ was then found to correlate these values linearly. As a result the exponents -0.33 and -0.70 were obtained for the Terumo TE10 and TH10 module, respectively. As shown with these exponents, the influence of Qw on the mass-transfer of coefficient of the outside liquid film was greater in the long hollow-fiber module than in the short one. The effects of a module's geometric factors, such as module diameter, fiber length, and packing ratio of fibers, on mass transfer and liquid flow characteristics should be studied in more detail to determine the optimum module construction.

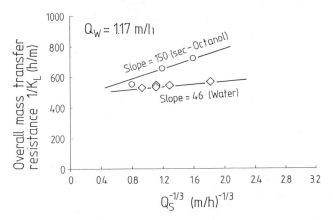

Figure 7 Applicability of Leveque's equation to hollow fibers.

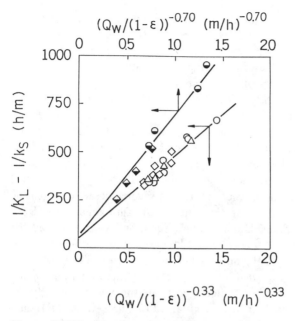

Figure 8 Wilson's plot of outside flow rate Qw to kL - ks for determining out-side liquid film resistance and membrane resistance. Keys the same as in Fig. 6.

The mass-transfer coefficients of the membranes were obtained from the intersection of these straight lines and the vertical axis at Qw = 0. Approximately the same values were obtained for the Terumo TE10 and TH10 modules. Each of these mass-transfer coefficients is summarized in Table 4. The solid and dotted lines in Fig. 6 show the values calculated from the mass-transfer coefficients in Table 4.

Kiani et al. (35) proposed a membrane-aided solvent extraction method using a microporous hydrophobic membrane, and extracted acetic acid from aqueous solution into either methyl isobutyl ketone or xylene. The pores of a flat poly-propylene film (Celgard 2400; membrane thickness 25.4 μm, porosity 38%)

Table 4 Ethanol Mass-Transfer Coefficients in Hollow-Fiber Modules

Module	k_L (m/h)	k_m (m/h)	k_s (m/h)
Terumo TE10	$2.33 \times 10^{-3} [Qw/(1 - \epsilon)]^{0.33}$	0.020	$50.0 Ds^{2/3} Qs^{1/3}$
Terumo TH10	$1.59 \times 10^{-3} [Qw/(1 - \epsilon)]^{0.70}$	0.015	$38.5 Ds^{2/3} Qs^{1/3}$

were filled with the solvent, and the solvent contacted the aqueous phase at the mouth of the pores. Acetic acid diffused into the pores from the aqueous phase and then into the bulk of the solvent. Using a flow-type extraction test cell for a flat membrane, they investigated the effects of aqueous and solvent flow rates on the overall mass transfer coefficient and estimated the mass-transfer coefficient k_m of the porous membrane filled with xylene to be 0.075 m/h at 25°C. This k_m value is more than three times higher than that of Cuprophan. It would probably be useful to improve the extraction performance, but pressure control in aqueous phase was essential to prevent the leakage of solvent from the pores.

D. Continuous Ethanol Production in Fermentation Systems Coupled with Perstraction

A schematic diagram of the fermentation system coupled with the perstraction process is shown in Fig. 9. A stirred fermentor equipped with a draft tube was used. A three-blade propeller placed in the draft tube discharged liquid upward,

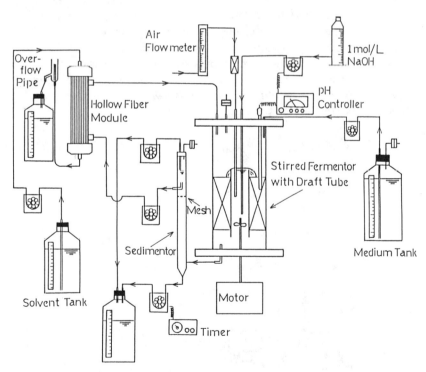

Figure 9 Experimental setup for continuous ethanol fermentation coupled with perstraction process.

so that the fermentation broth flowed up the draft tube and down the annulus between the draft tube and the fermentor wall. Gel beads immobilizing *S. cerevisiae* ATCC 26603, a flocculating yeast, were packed in the annulus, and the total working volume in the fermenter was adjusted to a given volume with medium. The gel beads contained 20 vol.% silanized silica gel, which acted as an adsorbent to trap solvent molecules coming into the gel beads. Since the density of gel beads containing active yeast is generally less than that of the fermentation broth, the gel beads float in normal stirred fermentor. In our fermentor the gel beads were well fluidized in the annulus, balancing their buoyancy force and the drag force of the fermentation broth. The fluidization was very effective in preventing channeling and blocking of the liquid flow in the gel bead bed caused by the accumulation of CO_2 gas and free cells growing in the fermentation broth after escape from the gel beads.

The fermentor was connected to a sedimentor. Because a flocculating yeast was used, the free cells settled in this sedimentor and the resulting sediment was removed periodically by a peristaltic pump. The supernatant in this sedimentor was circulated through a hollow-fiber module. Solvent was supplied countercurrently into the module, and ethanol in the fermentation broth was extracted through the membrane. The liquid level of the fermentor was controlled by a peristaltic pump connected to an overflow pipe in the sedimentor.

The extraction capacity of the hollow-fiber modules can be estimated by using the overall volumentric mass-transfer coefficient, which was determined from the mass-transfer coefficients in Table 4. When the hollow-fiber module is coupled with a continuous stirred fermentor as shown in Fig. 9, the following material balance of ethanol is obtained:

$$V_L \frac{d(C_{E/W})_1}{dt} = SQ_B[(C_{E/W})_2 - (C_{E/W})_1] - (F + f)(C_{E/W})_1 + V_G P_G \quad (7)$$

$$\frac{\partial(C_{E/W})}{\partial t} = -Q_B \frac{\partial(C_{E/W})}{\partial Z} - K_L a(mC_{E/W} - C_{E/S}) \quad (8)$$

$$\frac{\partial(C_{E/S})}{\partial t} = Q_S \frac{\partial(C_{E/S})}{\partial Z} + K_L a(mC_{E/W} - C_{E/S}) \quad (9)$$

At a steady state, this equation becomes

$$(C_{E/W})_1 = \frac{V_G P_G}{SQ_B(1 - \beta) + (F + f)} \quad (10)$$

$$(C_{E/S})_1 = (C_{E/W})_1 \frac{Q_B}{Q_S} (1 - \beta) \quad (11)$$

where V_G and V_L are the volume of gel beads and fermentation broth, Q_S and Q_B are the flow rates of solvent and circulated broth, C_S and C_B are the ethanol concentrations in solvent and broth, and F and f are the volumetric flow rates of fresh medium and 1 N NaOH for pH control. P_G is the ethanol productivity of the gel beads, and S is the cross-sectional area of the module. β is C_{B2}/C_{B1} as given by Eq (3). Subscript 1 indicates the broth inlet or solvent outlet of the module, and subscript 2 indicates the broth outlet or solvent inlet.

In the following experiments on continuous ethanol production, we decided to use tri-*n*-butylphosphate as the extractant, based on the results of previous ethanol perstraction experiments (see Fig. 6) and its weak solvent toxicity (see Fig. 3). Operating conditions for perstraction with this solvent were determined by referring to the estimated ethanol concentrations from Eqs. (10) and (11).

Figure 10 shows an example of continuous ethanol production in the system coupled with perstraction using the TE10 module. The volumetric flow rates of solvent and circulated broth were 0.57 and 1.7 L/h, respectively. Under these operating conditions the overall volumetric mass-transfer coefficient was estimated as 10.4 L/h. Fresh gel beads of 0.2 L were introduced into the fermentor

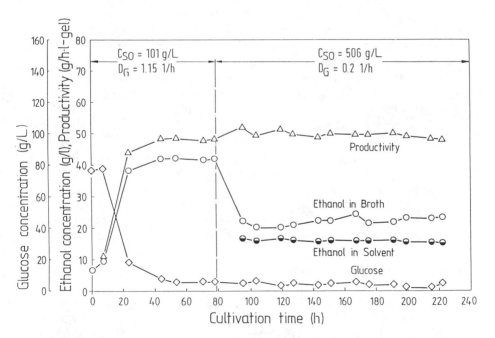

Figure 10 Continuous ethanol production coupled with perstraction. Operating conditions: immobilized cells, *S. cerevisiae* ATCC 26603 immobilized together with 20 vol.% silica gel; extractant, tri-*n*-butylphosphate; module, Terumo TE10; flow rate of extractant, 0.57 L/h; circulation flow rate of broth, 1.7 L/h; kLa, 10.4 1/h; gel bead volume, 0.2 L; broth volume, 1 L.

with a 1.0 L growth medium containing 100 g/L glucose. These gel beads were held in the medium for a day without mixing of broth. During this period, almost all the gel beads floated. The mixing of broth and feeding of a growth medium containing 101 g/L of glucose was then started. The dilution rate based on the volume of gel beads D_G was 1.15 L/h·L gel. This feeding continued for 80 h to confirm the ethanol productivity of the gel beads, and a productivity of 48 g/h·L gel, which was the maximum attainable by yeast (36), was obtained.

After confirming the maximum ethanol productivity, the dilution rate and glucose concentration of feed medium were changed to 0.2 L/h·L gel and 506 g/L, respectively, and perstraction started. The ethanol concentration in the fermentation broth decreased rather quickly and was kept at a low level, around 20 g/L. The calculated values of ethanol in the broth and the solvent were 20.6 and 15.5 g/L, respectively, and these values agreed fairly well with the measured values. The ethanol productivity, which was calculated from the amount of ethanol coming out of the fermentor and the hollow-fiber module, was the same as that in the 101 g/L glucose medium. This means that ethanol inhibition was completely removed by the perstraction and that silanized silica gel protected yeast cells from solvent toxicity. However, we had to stop the experiment after 220 h because the housing case of the module cracked. This trouble was caused by solvent leaking from the hollow fibers. The leaked solvent accumulated as relatively large droplets in the module. These droplets attached to the module wall and cracked the housing case.

Another result of continuous ethanol production is shown in Fig. 11. In this experiment the TH10 module, with a thicker membrane than that in the TE10 module, was used, and the volumetric flow rates of solvent and circulated fermentation broth were 0.125 and 1.7 L/h, respectively. The overall volumetric mass-transfer coefficient was estimated as 5.7 L/h. The conditioning of the gel beads continued for 90 h, and then the feed rate and glucose concentration of the medium were changed. The glucose concentration was increased from 110 to 520 g/L, and the feed rate was decreased from 1 to 0.2 L/h. A new steady state was attained after 180 h. The ethanol productivity in this steady state was the same as that in the growth medium of 110 g/L glucose, and the ethanol concentration in the fermentation broth and solvent agreed well with the estimated values from Eqs. (10) and (11) (67 g/L broth; 53 g/L solvent). The steady state continued for 340 h, but then the productivity began to decrease and the residual glucose concentration increased. This could only have resulted from the deactivation of immobilized yeast cells due to the decrease in the protection effectiveness of the silanized silica gel. Employment of a thicker membrane greatly reduced the leakage of solvent.

We succeeded in fermenting a 500 g/L glucose medium without ethanol inhibition, and also in reducing solvent requirement. The solvent requirement per consumed glucose, obtained from the volumetric flow rate of solvent and

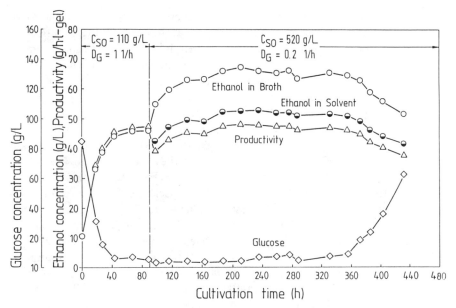

Figure 11 Continuous ethanol production coupled with perstraction process. Operating conditions: immobilized cells and extractant are same as in Fig. 10; module, Terumo TH10; flow rate of extractant, 0.125 L/h; circulation flow rate of broth, 1.7 L/h; kLa, 5.7 1/h; gel beads volume, 0.2 L; broth volume, 1 L.

glucose consumption rate, was 6.0 L solvent per kg glucose. This value is extremely small when compared with the result of Minier and Goma (43 L solvent per kg glucose) (8), who used nontoxic n-dodecanol with a poor ethanol distribution coefficient as the extractant. However, many problems remain, which should be solved to make the perstraction process more practical.

E. Problems of Perstraction Process and Countermeasures

In the previous sections we showed the usefulness of the perstraction process, but the following problems remain to be solved.

1. The protection effectiveness of the adsorbent in gel beads is lost rather quickly.
2. The solvent requirement is still high.
3. Research into ways to recover the ethanol from the solvent should be done.

1. Durability of Protection Effect of Adsorbent

As mentioned previously, the durability depends on the amount of adsorbent in the gel beads and on the concentration of solvent in the fermentation broth. It is

difficult to increase the adsorbent amount to more than 20 vol.% without sacrificing the solidity of the gel beads. Even if it were possible, the effectiveness
would be lost sooner or later when the adsorbent is saturated with solvent molecules. Therefore, we tried to reduce the load on the adsorbent in the gel beads
by eliminating solvent molecules in the circulating broth as much as possible just
after it had passed through the hollow fiber module.

One approach is the adsorption in a packed column. We made small alginate
gel beads containing only silanized silica gel by using an atomizer. This immobilization is for reducing the pressure drop in the packed column. The average
diameter of these gel beads was 1.1 mm, and they exhibited the Freundlich
isotherm against tri-n-butylphosphate as shown in Fig. 12. The breakthrough
curves of tri-n-butylphosphate and ethanol in a packed column (column diameter 25 mm, packed length 400 mm, packing ratio 0.7) containing the immobilized silanized silica gel beads are shown in Fig. 13. In these experiments an
aqueous ethanol solution of 50 g/L saturated with tri-n-butylphosphate was
supplied at a linear velocity of 74 cm/h. The ethanol was immediately saturated,
but it took a rather long time for tri-n-butylphosphate to attain the break
point. Using these experimental results, we estimated the overall volumetric
mass-transfer coefficient of tri-n-butylphosphate into the gel beads to be 91
L/h. After obtaining the overall volumetric mass-transfer coefficient, we can
estimate the break point under any operating conditions according to the

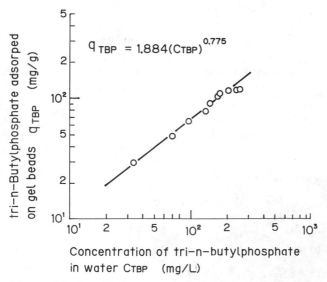

Figure 12 Equilibrium adsorption isothermal of tri-n-butylphosphate for Ca
alginate gel beads containing silanized silica gel.

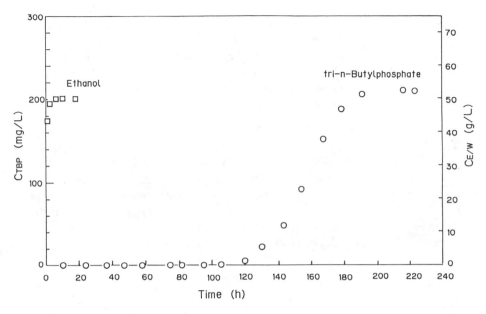

Figure 13 Breakthrough curve of tri-*n*-butylphosphate and ethanol in a column packed with Ca alginate gel beads containing silanized silica gel.

method proposed by Hashimoto and Miura (37). We can therefore exchange the packed column for a new one before it attains the break point, and by this method the durability of the protection effect may be greatly prolonged. Solvent extraction in a spray column, for example, would also be applicable to elimination of the solvent molecules in the circulated broth and in prolonging the durability of the protection effect. In this case it would be necessary to use a nontoxic solvent with a large density difference and a quite low ethanol distribution coefficient, such as *n*-paraffin (38).

To operate the perstraction process and maintain stability, we must prevent solvent leakage from hollow fibers and cracking of the housing case. In our experiments we used hollow-fiber modules that were developed for artificial kidneys. The membrane thickness of the hollow fiber for artificial kidneys is becoming thinner as development advances to increase the mass-transfer rate and decrease the size of the module. However, in a perstraction process using organic solvents, stable operation should take precedence over miniaturization of the hollow-fiber module; we therefore produced a hollow-fiber module comprised of a thicker membrane and a housing case that is stable against solvents. We selected a hollow fiber made of polyvinyl alcohol (350 μm ID, 250 μm OD; Kuraray Co. Ltd., Japan) and a module with a housing case made of glass of the type used in industrial dialyzers. A large-scale module with a constant area of 100 m² is commercially available. We tested the extraction performance of this hollow-fiber module and found almost the same overall mass-transfer coefficient as that of the Terumo TH10 module despite its high membrane thickness. In this polyvinyl alcohol hollow-fiber module, leakage of tri-*n*-butylphosphate and cracks in housing case were prevented.

2. Reduction of Solvent Requirement

The ethanol distribution coefficient determines the solvent quantity necessary to extract ethanol. Since the affinity between ethanol and water is strong, it is not easy to find a suitable solvent with a high distribution coefficient. Phenol derivatives, such as *o*-isopropylphenol and *o-tert*-butylphenol, have the highest ($m = 1.4$) of the solvents investigated so far. However, we should make efforts to find a more suitable solvent by using a systematic method based on thermodynamic relations, such as those of ASOG (39) and UNIFAC (40).

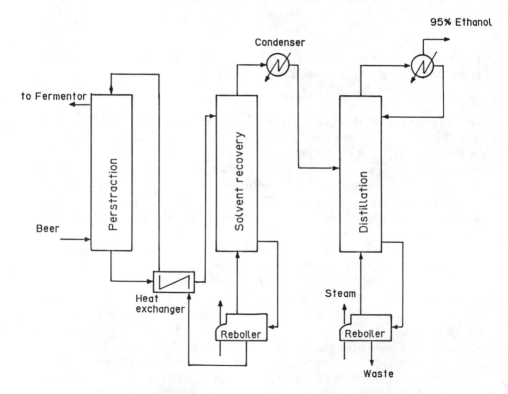

Figure 14 Ethanol purification system: coupled perstraction and distillation.

3. Recovery of Ethanol from Solvent

We have not yet studied this problem but are of the mind that distillation as shown in Fig. 14 is the most practical method to recover the solvent and to concentrate ethanol. In the solvent recovery process a distillation under reduced pressure may also be applicable. In the example shown in Fig. 11, 53 g/L of ethanol was contained in the solvent, and water was almost at the saturation concentration of 64.5 g/L (see Table 2). Therefore, the ethanol concentration becomes 45 wt.% on a solvent-free basis. This concentration is the same as that from a beer in the conventional two-column ethanol-water distillation system (41). In a system with a solvent recycled between the hollow-fiber module and solvent recovery column, we need not completely remove the solutes from the solvent. We want to leave as much water as possible in the solvent. To determine the optimum operating conditions, research into the binary vapor-liquid equilibrium in a solvent with a high boiling point is necessary.

III. PERSTRACTION USING A LIQUID MEMBRANE

Membrane separation processes are generally considered less energy intensive, and the use of membranes for separating and concentrating ethanol from fermentation broths has been suggested. Since reverse osmosis does not involve energy-intensive processes, such as vaporization and condensation, it showed the most promise as a separation method for aqueous ethanol solution (11–13). However, because of the high osmotic pressure of an ethanol-water mixture, the concentration of ethanol obtained was limited to 20 wt.% (42).

On the other hand, in pervaporation as proposed by Binning et al. (43), this limitation of osmotic pressure imposed on reverse osmosis is avoided by maintaining the permeate below its saturation vapor pressure. The vapor permeate is usually removed by vacuum pumping, but gas sweeping (44) or condensating on a cold wall (45) can also be used. Pervaporation includes the processes of vaporization and condensation, so the energy efficiency of alcohol recovery is less than that in reverse osmosis. However, more concentrated permeate can be expected. Hoover and Hwang (15) and Kimura and Nomura (16) applied pervaporation with a silicone rubber membrane to the separation of ethanol, isopropanol, n-propanol, and methanol from their aqueous solutions. They found that the silicone rubber membrane permeated these alcohols preferentially and the ethanol separation factor for aqueous ethanol solutions less than 20 wt.% was about 9. This separation factor means that a biomass ethanol solution of 7 wt.% can be concentrated up to 40 wt.%. A number of researchers are trying to develop a solid membrane superior to the silicone rubber membrane in both selectivity and permeability. Despite their efforts a really efficient solid membrane reamins elusive (46).

Extraction processes with liquid membranes are also attractive. They have been applied to the separation of metal ions and hydrocarbons, as well as wastewater treatment for the remvoal of pollutants, such as phenol and ammonia (47). The broad applicability of this extraction process may be attributed to the ease in changing the membrane characteristics.

We combined solvent extraction with membrane separation to make a new perstraction process for separating volatile products from fermentation broths (22). This new process is based on pervaporation using a liquid membrane supported with a microporous hydrophobic membrane. In this section I introduce basic studies of this process as they were carried out primarily in the separation of butanol.

A. Selection of Liquid Membrane

A microporous polypropylene flat sheet, Celgard 2500 (Celanes plastic, Charlotte, NC), was used as a support material for liquid membranes. This hydrophobic film has the following physical properties; thickness 25 μm, porosity 45%, maximum size of slender pore 0.04 $\times$ 0.4 μm, and critical surface tension (the maximum surface tension of the liquid that can pass through the pore of this membrane) 35 mN/m. This flat sheet was immersed in a petri dish containing a solvent. After 24 h the membrane, which had absorbed the solvent into its pores by capillary force, was taken out of the petri dish and pressed between paper tissues to remove excess solvent on the membrane surface.

Table 5 shows the physical properties of the solvents used in our experiments; boiling point (BP), density ρ, viscosity μ, surface tension of solvent (ST), interfacial tension between water and solvent (IT), distribution coefficient of butanol, solubility of water in solvent Sw/s, and solubility of solvent in water Ss/w. The distribution coefficients of butanol are much higher than those of ethanol.

Using the liquid membranes prepared with the 10 solvents in Table 5, the stability of a liquid membrane under vacuum was investigated. Pervaporation experiments of a butanol solution with a weight fraction of $X_B = 0.01$ were carried out at a downstream pressure of 0.133 kPa in the apparatus shown in Fig. 15. The permation cell was made of stainless steel.

A membrane set on a stainless mesh disk separated the cell into two compartments. The upper compartment had a capacity of about 100 ml, and the effective membrane area in contact with a feed solution was 25.5 cm^2. It was equipped with a heating jacket to keep the feed solution at 30°C. The feed solution was mixed by a magnetic stirrer and circulated by a peristaltic pump between the permeation cell and a feed tank to prevent any change in feed concentration. The outlet joint from the lower compartment was connected to traps in a methanol bath cooled to -20°C by an electric cooling unit. This

Table 5 Physical Properties of Organic Solvents for Preparing Liquid Membranes and Stability of Liquid Membranes Under Vacuum

Solvent	BP (°C)	ρ (kg/m^3)	μ (mPas)	ST (mN/m)	IT (mN/m)	Distribution coefficient	Solubility $S_{w/s}$ (wt.%)	$S_{s/w}$ (wt.%)	Separation factor (–)	Permeation flux (kg/m^2.h)
Oleyl alcohol	207 at 1.7 kPa	850	26.0	31.6	13.6	3.8	1.14	0.0019	180	0.080
n-Decyl alcohol	231	830	11.3	27.6	8.0	5.7	3.77	0.0727	–	–
2-Ethylexyl alcohol	185	830	8.8	26.0	<5	7.6	2.77	0.0067	–	–
n-undecyl alcohol	146 at 4.0 kPa	830	12.4	28.4	8.0	6.0	3.58	0.0033	–	–
1-Tridecyl alcohol	155 at 2.0 kPa	820	23.0	27.8	5.5	5.1	1.78	0.0024	–	–
d,l-n-Butyl-phthalate	339	1050	15.4	35.2	19.4	1.8	0.631	0.0040	90	0.112
Tri-n-butyl-phosphate	266	980	5.5	27.8	6.5	8.8	7.04	0.0421	–	–
n-Butyl-n-caprylate	241	880	3.5	26.8	26.0	1.6	0.36	0.0049	–	–
2-Ethylhexyl-phosphate	209 at 1.3kPa	965	162	27.2	<5	8.0	1.28	0.0010	–	–
Tricresyl-phosphate	250 at 0.5 kPa	1170	46.5	40.4	16.0	2.3	0.28	0.0020	105	0.055

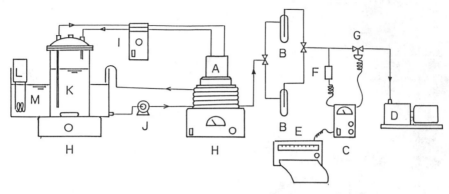

Figure 15 Experimental setup for pervaporation: (A) permeation cell; (B) cold trap; (C) Pirani gage; (D) vacuum pump; (E) recorder; (F) pressure sensor; (G) solenoid valve; (H) magnetic stirrer; (I) peristaltic pump; (J) liquid circulation pump; (K) feed tank; (L) heater; (M) water bath.

cooling method was replaced by liquid nitrogen when the required temperature was lower than –100°C. The pressure on the downstream side was maintained constant by using a solenoid valve controlled by a Pirani gage. The measured permeation fluxes and separation factors a defined by Eq. (12) are summarized in Table 5. The dash in the columns headed separation factor and permeation flux indicates that the liquid membrane was partially or completely broken under the vacuum of 0.133 kPa and stable operation was impossible.

$$a = \frac{Y/(1 - Y)}{X/(1 - X)} \tag{12}$$

where Y and X are the weight fractions of the components that preferentially permeate in the permeate and feed solution, respectively. As can be seen from Table 5, all the solvents that have an interfacial tension less than 10 mN/m were lost from the pores of the support membrane under vacuum. The experimental result with 2-ethylhexylphosphate shows that the viscosity of the solvent was not the major factor for stabilizing the supported liquid membrane. The liquid membranes prepared with oleyl alcohol, di-n-butylphthalate, and tricresylphosphate proved to be stable. Because the unstable liquid membranes were prepared with solvents having an interfacial tension less than 10 mN/m, interfacial tension may have been a criterion for the first choice of a solvent to make a stable liquid membrane. However, we found an exception, that is, the liquid membrane prepared with n-butyl-n-caprylate. Despite its high interfacial tension the liquid membrane was unstable under vacuum. The viscosity of n-butyl-n-caprylate is very low compared with that of the three stable solvents. Viscosity was not a major factor in the stabilization of liquid membranes, as mentioned previously, but it may partially contribute to the stabilization.

We selected a liquid membrane prepared with oleyl alcohol. This selection was based on the separation factor rather than the permeation of flux, because the permeation rate can be increased rather easily by using a hollow-fiber module with a high contact area.

B. Separation Performance of an Oleyl Alcohol Liquid Membrane

Pervaporation experiments of dilute aqueous solutions of butanol, ethanol, isopropanol, and acetone were carried out at a downstream pressure of 0.133 kPa. Figure 16 shows the results of pervaporation of an aqueous butanol solution with the oleyl alcohol liquid membrane. The butanol concentration in the permeate Y_B agreed fairly well with the solid line obtained from Eq. (12), assuming that the separation factor a was 180. This separation factor means that a biomass butanol solution of around 4 g/L can be concentrated up to 100 times. The total permeation flux J for the aqueous butanol solutions in the concentration region lower than $X_B = 0.015$ increased linearly with increasing

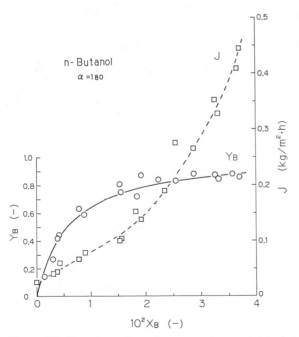

Figure 16 Pervaporation of aqueous butanol solutions.

butanol concentration in the feed solution. Above this permeant concentration it showed an accelerative increase, and the liquid membrane became unstable when the permeant concentration approached 0.05.

Figures 17 and 18 compare the separation performance of the oleyl alcohol liquid membrane with that of other solid membranes. As can be seen from Fig. 17, the butanol selectivity of the liquid membrane was greatly superior to that of silicone rubber membrane and polypropylene hollow fiber. Silicone rubber membrane is well known for its high selectivity for alcohols, but the separation factor was about 70. Ohya et al. (19) tried to concentrate an aqueous butanol solution by pervaporation with a porous polypropylene hollow-fiber membrane (outer diameter 262 μm, thickness 20.6 μm, pore size 0.083 μm, porosity 0.44) that had properties similar to our support membrane. The separation factor was only 5 and did not exceed even the vapor-liquid equilibrium for an aqueous butanol solution at a constant pressure of 0.1013 MPa, shown by the dotted line marked V.L.E. Figure 18 shows the butanol permeability of the three membranes. The liquid membrane also emerged as excellent in permeability. The poor permeability of the silicone rubber membrane is due to its thickness of 180 μm. Production of a very thin silicone rubber membrane has been tried by using plasma polymerization. If this becomes available, its permeability will be highly improved.

Figure 19 shows the pervaporation results of aqueous ethanol solutions. In these experiments two trapping methods were used. One was trapping using a single trap cooled by liquid nitrogen, and the other used two traps connected

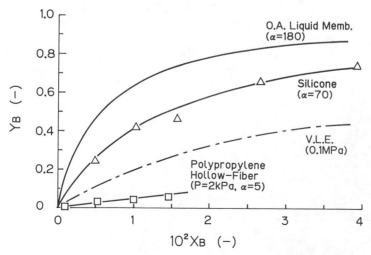

Figure 17 Comparison of butanol selectivity between oleyl alcohol liquid membrane and solid membranes.

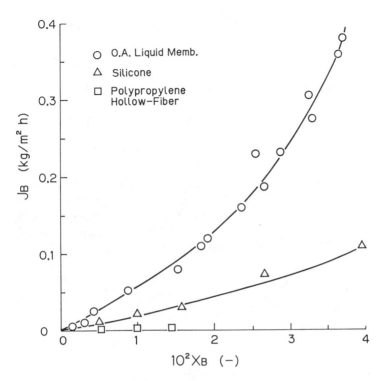

Figure 18 Comparison of butanol permeability between oleyl alcohol liquid membrane and solid membranes.

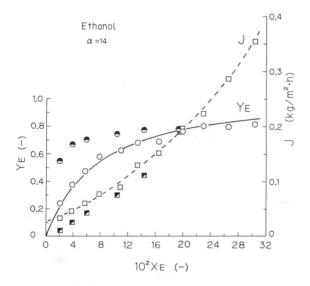

Figure 19 Pervaporation of aqueous ethanol solution. Half-black keys show the results obtained in the double-trap method.

in series, one cooled at -20°C and the other by liquid nitrogen, respectively. The results obtained in the single-trap method are shown by white keys; the results in the double-trap method are shown by half-black keys (Fig. 19). The separation factor in the single-trap method was 14, and the permeation flux showed a nonlinear increase as the ethanol concentration in the feed solution increased. In the double-trap method the first trap cooled at -20°C trapped only water, and as a result the ethanol concentration in the permeate collected in the second trap increased. The effect of this partial condensation appeared at an ethanol feed concentration of around X_E = 0.16 and became more marked as the ethanol concentration in the feed solution fell. The separation factor was about 37 for the aqueous ethanol solution, with X_E = 0.06. If this new separation process were applied to an ethanol fermentation, we could obtain more concentrated ethanol than in a perstraction process using the hydrophilic hollow-fiber membrane. The ethanol selectivity of the liquid membrane is compared with that of a silicone rubber membrane (16) and a porous polypropylene hollow fiber (17) in Fig. 20. The separation factor a of the liquid membrane is higher than that of the silicone rubber membrane (a = 9). By comparing the results between the liquid membrane and the porous polypropylene hollow fiber shown in Figs. 17 and 20, it is clear that oleyl alcohol soaked into the pores of polypropylene is quite effective in increasing the selectivity for volatile organic materials.

Figures 21 and 22 show the separation performance of the liquid membrane for acetone and isopropanol, respectively. Despite that oleyl alcohol was selected mainly for the separation of butanol, the liquid membrane also showed rather high selectivity for both acetone and isopropanol. Therefore, this liquid membrane could be considered for application in both acetone-butanol and isopropanol-butanol fermentation.

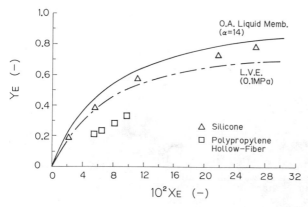

Figure 20 Comparison of ethanol selectivity between oleyl alcohol liquid membrane and solid membrane.

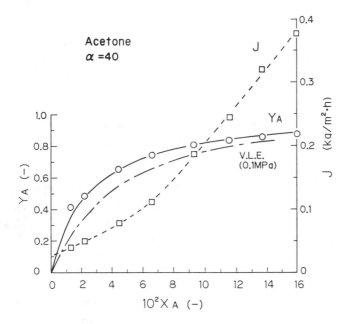

Figure 21 Pervaporation of aqueous acetone solutions.

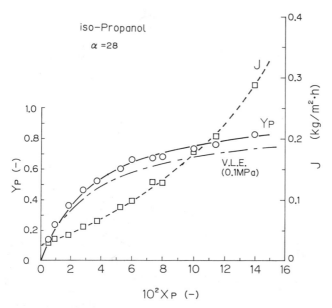

Figure 22 Pervaporation of isopropanol aqueous solutions.

C. Stability of Oleyl Alcohol Liquid Membrane

As shown in Fig. 16, pervaporation was successfully carried out for an aqueous butanol solution with a butanol concentration lower than X_B = 0.04, but the oleyl alcohol liquid membrane became unstable for solutions higher than X_B = 0.05. This liquid membrane was also applicable to more concentrated aqeuous solutions of ethanol, acetone, and isopropanol, as shown in Figs. 19, 21, and 22. This means that the stability of the liquid membrane was influenced by both the component and the concentration of feed solution. To investigate the factors affecting the stability of the liquid membrane, measurements of surface and interfacial tension were carried out in a system in which the solute was distributed between water and oleyl alcohol. From these investigations we found that the surface tension of the water was the most important factor affecting the stability of the liquid membrane. As shown in Fig. 23, the surface tension σ_W of the water containing butanol dropped markedly with the increase in butanol concentration $C_{B/W}$ and attained the critical tension of the support membrane (35 mN/m) at a concentration of $C_{B/W}$ = 50 g/L. This concentration almost coincided with the critical conccentration at which the liquid membrane became unstable. For the surface tension of the water containing the other solutes, a value more than 35 mN/m was maintained in a concentration region lower than 200 g/L. From these results we concluded that the liquid membrane was stable as long as the surface tension of feed solutions was higher than the critical surface tension of the support membrane. If a porous Teflon membrane with a

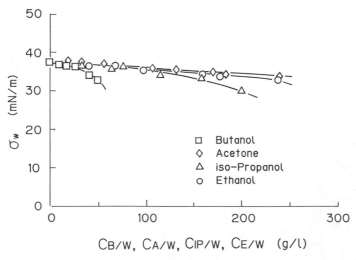

Figure 23 Influence of solute concentration on surface tension of aqueous phase.

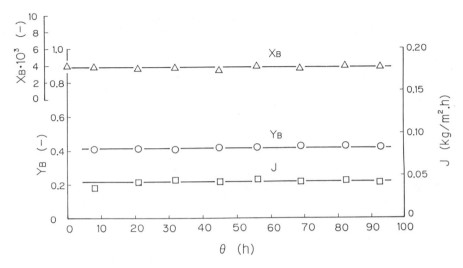

Figure 24 Stability of oleyl alcohol liquid membrane in continuous pervaporation of aqueous butanol solution.

critical surface tension of 29 mN/m is used as a support membrane, the concentration range to which the liquid membrane is applicable should be expanded more.

We also checked the stability of the liquid membrane for continuous long-term usage as shown in Fig. 24. In this experiment an amount corresponding to the amount of butanol permeated was supplied into the feed tank at every sampling time to keep the butanol concentration in the feed solution constant. As can be seen from Fig. 24, the stability of the liquid membrane was satisfactory for dilute aqueous butanol solutions corresponding to a butanol fermentation broth. However, if a pervaporation system using the liquid membrane was coupled with a continuous butanol fermentation and the liquid membrane contacted large amounts of the fermentation broth, the liquid membrane became thinner. The solubility of oleyl alcohol in water is very low, as shown in Table 5, but the oleyl alcohol in the support membrane still disappears little by little. This problem could be prevented by previously saturating the broth with oleyl alcohol. Ishii et al. (10) applied oleyl alcohol to make up a extractive fermentation system of acetone-butanol fermentation and found that oleyl alcohol was not toxic against *Clostridium acetobutylicum*. This indicates that it is possible to add a small amount of oleyl alcohol to a feed medium to prolong the life of the liquid membrane. In this case a fermentation system with a high feed concentration and a low dilution rate is most suitable to minimize the expenditure of oleyl alcohol.

D. Energy-Saving Effect of Perstraction Using Oleyl Alcohol Liquid Membrane

Such alcohols as ethanol and butanol produced from renewable biomass are attracting a great deal of attention as a promising feedstock for the chemical industry as well as an energy source (48). However, the net energy ratio of biomass alcohol, defined as the ratio of the total energy content of the useful products to the total agricultural and process energy inputs, has been argued against as unfavorable (49). Process energy inputs include energy for raw material processing, fermentation, product recovery, and effluent treatment. In conventional ethanol production processes over half this process energy requirement is for the steam used in the distillation of ethanol from the dilute fermentation broth. This is especially the case with butanol fermentation, in which the final butanol concentration in the batch culture is very low (around 0.5 wt.%) because of its serious product inhibition. The energy requirement for butanol purification by distillation must therefore be much higher than that of ethanol.

Assuming a simple distillation system consisting of a decanter, a butanol column, and an aqueous column, we calculated the energy requirement for butanol purification. As shown in Fig. 25, a beer containing 0.5 wt.% butanol is preheated up to the boiling point of azeotrope (92.7°C) and then fed into the decanter. In this decanter the beer separates into two liquid phases: a butanol-rich

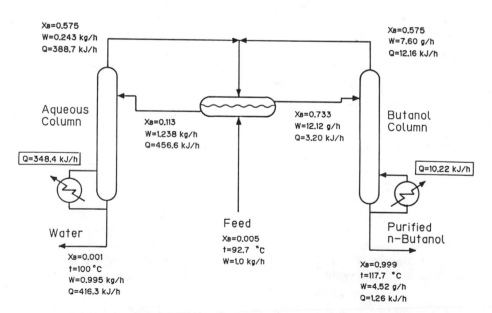

Figure 25 Mass and heat balance in butanol purification by distillation.

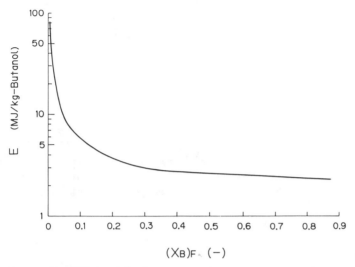

Figure 26 Effect of feed concentration on energy requirement of butanol purification by distillation.

phase with X_B = 0.733 and a water-rich phase with X_B = 0.133. The liquid in the butanol-rich phase is supplied into the butanol column, and a purified butanol with X_B = 0.999 is obtained as the bottom product. Water containing 0.1 wt.% butanol is discharged from the bottom of the aqeuous column. To obtain purified butanol at a rate of 4.52 g/h, energy supplements of 10.22 and 348.4 kJ/h are necessary in the butanol and aqueous columns, respectively. As a result, an energy expenditure of 79.5 MJ/kg butanol is required to purify a beer containing 0.5 wt.% butanol. Figure 26 shows the effect of the feed concentration $(X_B)_F$ on the energy requirement E for butanol purification by distillation. The energy requirement for distilling to 99.9 wt.% butanol falls sharply as the feed concentration increases to about $(X_B)_F$ = 0.4. Thereafter the energy requirement is virtually independent of the feed concentration. The energy requirement for purifying a beer containing 40 wt.% butanol is about 4% of that of a beer containing 0.5 wt.% butanol. the difference is quite significant. When pervaporation using the oleyl alcohol liquid membrane is applied to a beer containing 0.5 wt.% butanol, it can be concentrated to 45 wt.%, as shown in Fig. 16. The energy required in the pervaporation process was calculated as 4.67 MJ/kg butanol (50). Therefore if the pervaporation is employed as a pretreatment for the purification of butanol by distillation, the total energy requirement is 7.67 MJ/kg butanol. This energy requirement is still one-tenth that required for distillation. From these calculation results pervaporation using a membrane with both high selectivity and high permeability could be expected to bring about a large energy savings in a butanol purification process. We are also considering another butanol purification system by using membrane separations without distillation. We will report on the details of these separation systems in a forthcoming article (51).

In ethanol distillation as well, the feed concentration seriously affects its energy requirement. The energy requirement of distilling to 95 wt.% fell sharply as the feed concentration increased to about 7 wt.%, and thereafter it became virtually independent of the feed concentration (49). If this calculation is correct, it is impossible to find any merit for energy saving in the ethanol purification process even when pervaporation using the oleyl alcohol liquid membrane is employed, because 7 wt.% ethanol is easily obtained in a conventional fermentation system. However, a concentrated feed means less distillation waste and therefore lower waste treatment costs. As Murphy et al. explained (24), the energy requirement for wastewater treatment occupies a rather large portion of the process energy input for biomass ethanol production. Since our new separation process has good performance in concentrating ethanol, as shown in Fig. 19, it should contribute not only to the elimination of ethanol inhibition but also to saving the energy necessary for wastewater treatment.

IV. CONCLUSION

As mentioned previously, the validity of biomass alcohols is dependent on the development of a low-cost and energy-efficient recovery process. Solvent extraction is one of the most promising methods, because it has been established as a unit operation and employed on a large scale in many chemical industries. However, it has several problems that should be solved before it is applied to fermentation. One of them is solvent toxicity to microorganisms. I presented a method to protect microorganisms from solvent toxicity by using both a membrane and cells immobilized together with an adsorbent that has a strong affinity to solvents. Another problem is how to reduce the solvent amount required in the separation of alcohols from a fermentation broth. We overcame this problem by using a liquid membrane supported with a porous hydrophobic membrane under vacuum. In this case the solvent is required simply to fill the pores of the support membrane, and so the amount is extremely reduced. The oleyl alcohol liquid membrane showed excellent separation performance. However, it is not sufficient to purify alcohols from fermentation broth using only this membrane. In the present situation additional employment of distillation is essential to produce pure alcohols. Even so, we can achieve a rather high energy saving. Moreover, we should pay attention to the fact that the characteristics of liquid membranes can be easily changed by varying the type of solvent. We could also introduce a good carrier into the solvent to enhance its selectivity and permeability. When we succeed in greatly improving the quality of the solvent using such a carrier, it will be possible to replace distillation with a membrane separation technique that is less energy intensive.

V. NOMENCLATURE

a   specific contact area $(1/m)$

$C_{A/W}$ acetone concentration in aqueous phase (g/L)

$C_{B/W}$ butanol concentration in aqueous phase (g/L)

$C_{E/S}$ ethanol concentration in solvent phase (g/L or mol/L)

$C_{E/W}$ ethanol concentration in aqueous phase (g/L or mol/L)

C_G glucose concentration in fermentation broth (g/L)

$C_{IP/W}$ isopropanol concentration in aqueous phase (g/L)

C_{SO} glucose concentration in feed medium (g/L)

C_{TBP} concentration of tri-n-butylphosphate in water (mg/L)

D   diffusion coefficient of ethanol in extractant (m^2/h)

D_G dilution rate based on volume of gel beads (1/L gel·h)

d   inner diameter of hollow fiber (m)

E   energy requirement in distillation of butanol (MJ/kg butanol)

F   volumetric flow rate of fresh medium (L/h)

f   volumetric flow rate of 1 N NaOH solution (L/h)

J   total permeation flux $(kg/m^2 \cdot h)$

K_{La} overall volumetric mass-transfer coefficient (L/h)

k_L mass-transfer coefficient of outside liquid film of hollow fibers (m/h)

k_m mass-transfer coefficient of hollow fiber membrane (m/h)

k_s mass-transfer coefficient of inside liquid film of hollow fibers (m/h)

m   distribution coefficient

P   downstream pressure (kPa)

P_G ethanol productivity of immobilized cells (g/L gel·h)

Q   rate of energy supply (kJ/h)

Q_B superficial velocity of fermentation broth (m/h)

Q_S superficial velocity of solvent (m/h)

Q_W superficial velocity of aqueous ethanol solution (m/h)

q_{TBP} quantity of tri-n-butylphosphate adsorbed into gel beads containing silanized silica gel (mg/g)

Re   Reynolds number

Sh   Sherwood number

S   molar flow rate of solvent per cross-sectional area of hollow fiber module $(mol/m^2 \cdot h)$

$S_{S/W}$ solublity of solvent in water (g/L or wt.%)

$S_{W/S}$ solubility of water in solvent (g/L or wt.%)

t   cultivation time (h) or temperature (°C)

V_G volume of gel (L)

V_L volume of fermentation broth (L)

W   molar flow rate of aqueous ethanol solution per cross-sectional area of hollow fiber module $(mol/m^2 \cdot h)$ or feed rate in distillation process (kg/h)

X   weight fraction of solute in feed solution

$(X_B)_F$ feed concentration of butanol supplied into distiller

x   mole fraction of ethanol in solvent phase

Y   weight fraction of solute in permeate

y   mole fraction of ethanol in aqueous phase

Z   length of hollow fiber (m)

Greek Letters

α separation factor defined by Eq. (12)

β relative ethanol concentration defined by Eq. (3)

ϵ ratio of cross-sectional area of hollow fibers to that of module

θ operation time of perstraction (h)

μ viscosity of solvent (mPa·s)

ρ density of solvent (kg/m^3)

σ_W surface tension of aqueous phase (mN/m)

Subscripts

A   acetone

B   butanol

E   ethanol

P   isopropanol

ACKNOWLEDGMENT

The author is grateful to Prof. H. Märkl (Technische Universität Hamburg-Harburg) for giving me the opportunity to perform a part of this work in his laboratory and also thanks to Prof. H.Kataoka (University of Tsukuba) and Mr. M. A. Cullen for their valuable discussion and helpful suggestions.

REFERENCES

1. Ramalingham, A., and Finn, R. K., The vacuferm process: A new approach to fermentation alcohol, *Biotehcnol. Bioeng.* 19:583–489 (1977).
2. Cysewski, G. R., and Wilke, C. R., Rapid ethanol fermentation using vacuum and cell recycle, *Biotechnol. Bioeng.* 19:1125–1143 (1977).
3. Pitt, W. W., Jr., Haag, G. L., and Lee, D. D., Recovery of ethanol from fermentation broths using selective sorption-desorption, *Biotechnol. Bioeng.* 25:123–131 (1983).

4. Walsh, P. K., Liu, C. P., Findley, M. E., Liapis, A. I., and Siehr, D. J., Ethanol separation from water in a two-stage adsorption process, *Biotechnol. Bioeng. Symp.* 13:629–647 (1983).

5. Malik, R. K., Ghosh, P., and Ghose, T. K., Ethanol separation by adsorption-desorption, *Biotechnol. Bioeng.* 25:2277–2282 (1983).

6. Larsson, M., Holst, O., and Mattiasson, B., Butanol fermentation using a selective adsorption for product recovery, in: Proceedings of the Third European Congress on Biotechnology, Munich, West Germany, Vol. II, pp. 313–316 (1984).

7. Dadgar, A. M., and Foutch, G. L., The evaluation of solvents for the recovery of *Clostridium* fermentation products by liquid-liquid extraction, in: Proceedings of the 7th Symposium on Biotechnology for Fuels and Chemicals, Gatlinburg, TN, May 14–17 (1985).

8. Minier, M., and Goma, G., Ethanol production by extractive fermentation, *Biotechnol. Bioeng.* 24:1565–1579 (1982).

9. Honda, H., Taya, M., and Kobayashi, T., Ethanol fermentation associated with solvent extraction using immobilized growing cells of *Saccharomyces cerevisiae* and its lactose-fermentable fusant, *J. Chem. Eng. Japan* 19:268–273 (1986).

10. Ishi, S., Taya, M., and Kobayashi, T., Production of butanol by *Clostridium acetobutylicum* in extractive fermentation system, *J. Chem. Eng. Japan* 18:125–130 (1985).

11. Ohya, H., Kazama, E., and Negishi, Y., Concentration of ethyl alcohol aqueous solution by reverse osmosis, *Kagaku Kogaku Ronbunshu* 7:372–376 (1981).

12. Huang, S. Y., and Ko, W. C., Reverse osmosis concentration of ethyl alcohol solution, *Membrane* 9:113–120 (1984).

13. Choudhury, J. P., and Ghosh, P., Separation of ethanol-water mixture by reverse osmosis, *Biotechnol. Bioeng.* 27:1081–1084 (1985).

14. Garcia, A., Iannotti, E. L., and Fischer, J. L., Butanol fermentation liquor product and separation by reverse osmosis, *Biotechnol. Bioeng.* 28:785–791 (1986).

15. Hoover, K. C., and Hwang, S. T., Pervaporation by a continuous membrane column, *J. Membr. Sci.* 10:253–271 (1982).

16. Kimura, S., and Nomura, T., Pervaporation of organic substrate water system with silicone rubber membrane, *Membrane* 8:177–183 (1983).

17. Ohya, H., Matsumoto, H., Negishi, Y., and Matsumoto, K., Concentration of ethanol from its aqueous solution by pervaporation using porous polypropylene hollow-fiber membrane, *Membrane* 11:231–238 (1986).

18. Groot, W. J., Schoutens, G. H., Van Beelen, P. N., Van den Oever, C. E., and Kossen, N. W. F., Increase of substrate conversion by pervaporation in the continuous butanol fermentation, *Biotechnol. Bioeng. Lett.* 6:789–792 (1984).

19. Ohya, H., Matsumoto, H., Negishi, Y., and Matsumoto, K., Concentration of acetone and *n*-butanol from its aqueous solutions by pervaporation

using porous polypropylene hollow-fiber membrane, *Membrane* 11:285–296 (1986).

20. Matsumura, M., and Märkl, H., Application of solvent extraction to ethanol fermentation, *Appl. Microbiol. Biotechnol.* 20:371–377 (1984).

21. Matsumura, M., and Märkl, H., Elimination of ethanol inhibition by perstraction, *Biotechnol. Bioeng.* 28:534–541 (1986).

22. Matsumura, M., and Kataoka, H., Separation of dilute aqueous butanol and acetone solutions by pervaporation through liquid membranes, Accepted in *Biotechnol. Bioeng.* 30:887–895 (1987).

23. Crabbe, P. G., Tse, C. W., and Munro, P. A., Effect of Microorganisms on rate of liquid extraction of ethanol from fermentation broths, *Biotechnol. Bioeng.*, 28:939–943 (1986).

24. Murphy, T. K., Blanch, H. W., and Wilke, C. R., Water recycling in extractive fermentation, *Proc. Biochem.*, November/December 6–10 (1982).

25. Roddy, J. W., Distribution of ethanol-water mixtures to organic liquids, *Ind. Eng. Chem., Process Des. Develop.* 20:104–108 (1981).

26. Rose, D., Yeasts for molasses alcohol, *Process Biochem.* March, 10–36 (1976).

27. Hayashida, S., Feng, D. D., and Honga, M., Function of the high concentration alcohol producing factor, *Agri. Biol. Chem.* (*Tokyo*) 38:2001–2006 (1974).

28. Viikari, L., Nybergh, P., and Linka, M., Hydrolysis of cellulose by *Trichoderma reesei* enzymes and simultaneous production of ethanol by *Zymomonas* sp., *Advan Biotechnol.* 2:137–142 (1980).

29. Roger, P. L., Lee, K. J., Skotnicki, M. L., and Tribe, D. E., Ethanol production by *Zymomonas mobilis*, *Advan. Biochem. Eng.* 23:37–84 (1982).

30. Pye, E. K., and Humphrey, A. E., The biological production of liquid fuels from biomass, Univ. Penn. Interim Report to U.S. DOE, June–August (1979).

31. Playne, M. J., and Smith, B. R., Toxicity of organic extraction reagents to anaerobic bacteria, *Biotechnol. Bioeng.* 25:1251–1265 (1983).

32. Yumoto, S., and Ohya, H., Studies of dialysis-membrane performance for various solutes and effect of flow rate on membrane performance, *Kagaku Kogaku Ronbunshue* 8:150–154 (1982).

33. Bird, R. B., Stewart, W. E., and Lightfoot, E. N., *Transport Phenomena* John Wiley & Sons, New York, p. 399 (1960).

34. Bohrer, M. P., Diffusional boundary layer resistance for membrane transport, *Ind. Eng. Chem., Fundam* 22:72–78 (1983).

35. Kiani, A., Bhave, R. R., and Sirkar, K. K., Solvent extraction with immobilized interfaces in a microporous hydrophobic membrane, *J. Membr. Sci.* 20:125–145 (1984).

36. Wada, M., Kato, J., and Chibata, I., Continuous production of ethanol using immobilized growing yeast cells, *Eur. J. Appl. Microbiol.* 10:275–287 (1980).

37. Hashimoto, K., and Miura, K., A simplified method to design fixed-bed adsorbers for the Freundlich isotherm, *J. Chem. Eng., Japan* 9:388–392 (1976).

38. Roddy, J. W., and Coleman, C. F., Distribution of ethanol-water to normal alkanes from C_6 to C_{16}, *Ind. Eng. Chem., Fundam.* 20:250–254 (1981).

39. Tochigi, K., and Kojima, K., Prediction of liquid-liquid equilibria by an analytical solutions of groups, *J. Chem. Eng., Japan* 10:61–63 (1977).

40. Macedo, E., Weidlich, U., Gmehling, J., and Rasmussen, P., Vapor-liquid equilibria by UNIFAC group contribution, revision and extension 3, *Ind. Eng. Che., Process Des. Develop.* 22:676–678 (1983).

41. Maiorella, B., Wilke, C. R., and Blanch, H. W., Alcohol production and recovery, *Advan. Biochem. Eng.* 20:43–88 (1981).

42. Metha, G., Comparison of membrane processes with distillation for alcohol/water separation, *J. Membr. Sci.* 12:1–26 (1982).

43. Binning, R. C., Lee, R. J., Jennings, J. F., Martin, E. C., Separation of liquid mixtures by permeation, *Ind. Eng. Chem.* 53:45–50 (1961).

44. Yuan, S., and Schwartzberg, W. G., Mass transfer resistance in cross membrane evaporation into air, *A.I.C.H.E. Symp. Ser.* 68:41–48 (1972).

45. Aptel, P., Challard, N., Cuny, J., and Neel, J., Application of the pervaporation process to separate azeotropic mixtures, *J. Membr. Sci.* 1:271–287 (1976).

46. Schissel, P., and Orth, R. A., Separation of ethanol-water mixtures by pervaporation through thin, composite membranes, *J. Membr. Sci.* 17:109–120 (1984).

47. Way, J. D., Noble, R. D., Flynn, T. M., and Sloan, E. D., Liquid membrane transport: A survey, *J. Membr. Sci.* 12:239–259 (1982).

48. Schoutens, G. H., Groot, J. W., and Hoebeek, J. B. W., Application of *iso*-propanol-butanol-ethanol mixtures as an engine fuel, *Process Biochem.*, February:30 (1986).

49. Essien, D., and Pyle, D. L., Energy conservation in ethanol production by fermentation, *Process Biochem.* August:31–37 (1983).

50. Matsumura, M., Kataoka, H., Sueki, M., and Araki, K., Separation and concentration of volatile fermentation products by pervaporation through oleyl alcohol liquid membrane, in: Proceedings of the Annual Meeting of Fermentation Technology in Japan, Osaka, Japan, p. 148 (1986).

51. Matsumura, M., Kataoka, H., Sueki, K., and Araki, K., Energy saving effect of pervaporaion using oleyl alcohol liquid membrane in butanol purification, *Bioprocess Eng.* 3:93–100 (1988).

6

Extractive Bioconversions with Nonaqueous Solvents

Steven R. Roffler, Theodore W. Randolph, Douglas A. Miller, Harvey W. Blanch, and John M. Prausnitz

University of California at Berkeley
Berkeley, California

I. INTRODUCTION

Extraction with nonaqueous solvents provides a straightforward method for the recovery of products from bioconversion reactions. The in situ extraction of products during a bioconversion may be desirable because (1) the products adversely affect the bioconversion, or (2) the environment in the bioreactor adversely affects the products. This chapter presents a summary of extractive fermentations and extractive enzymatic reactions.

II. EXTRACTIVE FERMENTATION

A. Background

In many fermentations, accumulation of products in the broth inhibits cell growth or product formation. For example, during the production of acetone and butanol by *Clostridium acetobutylicum*, cell growth is totally inhibited by butanol concentrations of only 10–15 g/L (1,2). The excretion of fermentation products into the broth may also alter the physical environment of the cells in some undesirable way. For example, during the production of polysaccharides by fermentation, broth viscosity is greatly increased by the accumulation of polymers in the broth, making it more difficult to provide the cells with an adequate supply of oxygen (3). By removing the inhibitory products, fermentation can be carried out for extended periods of time at increased productivity.

In some instances it is desirable to remove products from the broth in situ because the products are adversely affected by the fermentation environment. Products can be degraded by proteolytic enzymes in the broth or damaged by the high shear rates generated in agitated fermentors. In addition, it is sometimes necessary to remove intermediate products as they are formed to prevent further reaction.

Whether the products inhibit the growth of the cells or the fermentation environment is deleterious to the products, solvent and broth must be in direct contact so that the products can be extracted into the organic solvent. Figure 1 shows two basic ways that broth and solvent can be in contact in extractive fermentation. In the first method, extraction solvent is introduced directly into the fermentor. The solvent is dispersed into the broth by agitating the contents of the fermentor, thereby creating a large surface area for the transfer of products into the solvent. Broth and solvent are subsequently separated by gravity or centrifugation. Products extracted into the solvent are later recovered, and the regenerated solvent is recycled to the fermentor. Extracted broth exiting the

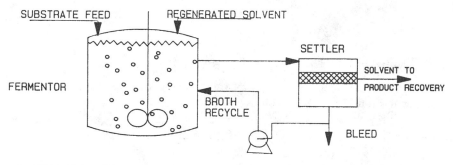

a) EXTRACTION WITH DIRECT SOLVENT ADDITION

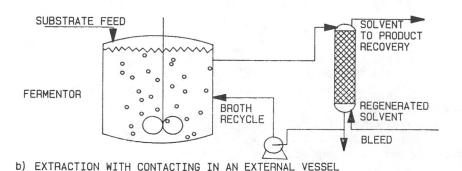

b) EXTRACTION WITH CONTACTING IN AN EXTERNAL VESSEL

Figure 1 Two basic modes of contacting solvent and broth.

settler can be recycled to the fermentor or discarded. In principle, the fermentor can be operated in batch, fed-batch, or continuous fashion and the solvent can be introduced into the fermentor either continuously or periodically.

Figure 1 also shows a second method of operating an extractive fermentation. Whole broth containing viable cells is cycled to an external extraction device where it is contacted with the extraction solvent. The extractor can be a single-stage contactor, such as an agitated tank, or a multistage device, such as an extraction column in which the extraction solvent and broth are contacted countercurrently. Extracted broth is recycled to the fermentor or discarded and the loaded extraction solvent is processed downstream in the product-recovery section.

Each of these two methods of extractive fermentation has advantages. Adding the solvent directly to the fermentor is simple and does not require an external extraction device. Direct addition of solvent to the fermentor, however, can lead to the formation of stable emulsions of organic solvent in the broth. High levels of agitation are required to disperse the extraction solvent in the broth. Small solvent drops formed near the impeller can be stabilized by surface-active components of the broth, inhibiting solvent-phase coalescence in the settler. Long settling times or mechanical centrifugation is then required to break the emulsion. Emulsion problems are reduced by contacting solvent and broth in an external extraction device. Levels of shear are generally more uniform in mechanically agitated extraction columns (4), and the formation of very small solvent drops is thus reduced. In addition, a more complete extraction can be obtained in a multistage contactor because only one equilibrium stage can be obtained when solvent is added directly to the fermentor. Solvent is therefore more efficiently used in an extraction column.

B. Solvent Selection

Regardless of whether solvent and broth are contacted in the fermentor or in an external extractor, the crucial decision is the selection of an extraction solvent. Important considerations for such selection are discussed here.

1. Solvent Biocompatibility

The largest constraint in selecting an extraction solvent is the requirement that the solvent be biocompatible. Extraction solvent and broth come into direct contact during extraction, and the extraction solvent must therefore not inhibit the growth of the cells. Identification of organic extractants that are nontoxic to growing cells has been the major problem that has so far prevented widespread application of extractive fermentation.

Several studies have investigated the biocompatibility of potential extraction solvents for extractive fermentation. Finn (5) examined the effect of straight-chain paraffin hydrocarbons on the growth of the yeast *Saccharomyces cerevisiae* and three bacteria. The results do not indicate whether the growth of the yeast cells depends on the concentration of hydrocarbon in solution or on the length of the hydrocarbon chain. Also, results obtained with gram-negative bacteria (*Serratia marcescens*) were different from those for gram-positive bacteria (*Sarcina flava* and *Bacillus subtilis*). Growth of the gram-positive bacteria was reduced by a factor of 2 when grown in broth saturated with tridecane; growth of the gram-negative bacteria was not affected to the same extent.

Several studies have reported the effects of organic solvents on the bacteria *C. acetobutylicum* used for the industrial production of acetone and butanol. Pye (6) found that the growth of these bacteria was not strongly inhibited by corn oil, olive oil, mineral oil, dibutyl ether, hexane, trichloroethylene, *n*-octanol, 2-octanol, decanol, butyloctyl phthalate, or butyl oleate. Hashimoto (7) reported that corn oil, butyloctyl phthalate, butyl oleate, and dibutylphthalate were nontoxic to *C. acetobutylicum* but hexane, *n*-octanol, and 2-octanol were toxic. In another study (8) it was found that all alcohols tested inhibit the growth of these bacteria. Alcohols tested included allyl phenol, nonyl phenol, amyl alcohol, heptyl alcohol, 2-ethyl hexanol, undecanol, 6-*tert*-butyl-2,4-xylenol, and tridecanol. Traxler et al. (9) found that hexanol, octanol, decanol, cyclohexanol, and 4-methyl cyclohexanol inhibited the growth of *C. acetobutylicum*, but hexadecanol and ethylcaproate were biocompatible with these cells. Roffler et al. (10) examined the toxicity of several alkanes, esters, and alcohols to this strain of bacteria. Biocompatible solvents included kerosene, cyclooctane, cyclohexane, dodecane, undecanone, nonane, benzyl benzoate, diethylphthalate, dibutylphthalate, dodecanol, and oleyl alcohol. Alkanes smaller than heptane, alcohols smaller than dodecanol, and most esters inhibited the growth of the cells to some degree. Ishii et al. (11) found that oxocol (branched-chained C_{14}-C_{15} alcohols), C_{16} Guerbet alcohol, oleyl alcohol, fine oxocol (branched-chained C_{18} alcohol), oleic acid, isosteric acid, Freon E, and octadecafluorodecalin were biocompatible with *C. acetobutylicum* (IAM 19012). In general, alkanes larger than hexane or heptane, alkyl phthalates, and high-molecular-weight esters were found to be biocompatible with *C. acetocutylicum*. Most researchers found that alcohols smaller than decanol or tetradecanol inhibit the growth of these cells.

Biocompatibility studies have been reported for solvent extraction for several other strains of bacteria and yeast. Minier and Goma (12) found that *S. cerevisiae* (UG5) was not inhibited by alcohols higher than dodecanol. Matsumara and Markel (13) found that *n*-octanol, 2-octanol, 2-ethyl-1-hexanol, 3-phenyl-1-propanol, tributylphosphate, and 2-ethyl-1-butanol inhibited the growth of several ethanol-producing microorganisms, but methyl crotonate and 2-ethyl-1,3-hexanediol had little effect on cell growth.

Datta (14) investigated the effects of several solvents on anaerobic acidogenic bacteria used to produce organic acids. No toxicity was observed when the broth was saturated with diesel oil, toluene, or amyl acetate with or without the addition of TOPO (trioctylphosphine oxide) or Alamine 336 (a tertiary amine).

Dave et al. (15) studied the effects of several extraction solvents used in the metal-refining industry on green algae and two cellulose-degrading bacteria, *Cellulomonas* sp. and *Sporocytophaga myxococcoide*. Table 1 lists the solvents tested. Primene JM–T, Adogen 283, Alamine 336, and Aliquat 336 were found to retard the growth of algae at concentrations between 0.1 and 0.5 mg/L. Adogen 383 and Amberlite LA-1 were only slightly less inhibitory. The other organic solvents were much less inhibitory. The bacteria were similarly affected by these extraction solvents, although they could tolerate higher concentrations of solvent.

Roffler (16) examined the effects of several classes of extraction solvents on *Lactobacillus delbreuckii* (NRRL-B445), a lactic acid-producing bacterium. Growth and product formation were not affected by the presence of the hydrocarbons dodecane, heptadecane, cyclooctane, kerosene, or cumene. The ketones tested, diisobutyl ketone, octanone, and undecanone, however, slowed or stopped growth of the cells. Phosphorus-based extractants, OPO (trioctylphosphine oxide), and tributylphosphate, also inhibited the growth of *L. delbreuckii* at saturation concentrations, although at 10% of saturation these solvents had no effect on the growth of the cells. Amines were also tested for their effect on the growth of *L. delbreuckii*. Table 2 shows the results of toxicity tests in which secondary and tertiary amines were added to the fermentation broth at two different concentrations, saturation and 10% of saturation. At 10% of saturation, only Adogen-283 (ditridecylamine), Adogen-381 (triisooctylamine), and Adogen-363 (trilaurel amine) had no affect on the rate of cell growth. Adogen-381, however, decreased the yield of lactic acid. When the fermentation broth was saturated with secondary or tertiary amines, growth of the cells was inhibited in all cases. Straight-chain amines appear to be more toxic than amines with branched alkyl groups, possibly because straight-chain alkyl groups may more easily enter the cell wall and disrupt cell functions. The biocompatibility of Aliquat-336, a quaternary amine, was also tested. Figure 2 shows the results of a continuous lactic acid fermentation in which Aliquat 336 was added to the broth. Aliquat-336 was found to be very toxic to *L. delbreuckii*: 0.025 g/L of Aliquat-336 reduced the growth of the cells below 0.2 h^{-1}. This concentration is only about 1% of the saturation concentration of Aliquat-336 in water.

Torma and Itzkovitch (17) studied the effects of several chelating solvents and amine extractants on the chemolithotropic bacteria *Thiobacillus ferrooxidans*. Table 3 lists the solvents tested. Inhibition of the cells increased in the order LIX 70 < LIX 73 < LIX 71 < UX 64N < LIX 65N < TBP ~ isodecanol ~ nonyl phenyl < LIX 63 ≪ D_2 EHPA ~ Kelex 100 < Kelex 120 ≪ Alamine 336 ~ Alamine 308 ~ Alamine 304 < Adogen 381 ~ Aliquat 336 < Adogen 364.

Table 1   Solvents Tested for Toxicity to Green Algae and Two Cellulose-Degrading Bacteria

Solvent	Structure	Main chemical components	Molecular weight
Primene JM-T	$M_2N-C_{20}M_{14}$	$C_{18}-C_{20}$ primary amine, branched	169–325
Amberlite and Adogen 283	$MN{\diagup}^{C_{13}H_{27}}_{\diagdown C_{13}H_{27}}$	High MW secondary amine	351–393
Adogen 383	$C_{13}H_{27}-N(H){\diagup}^{C_{13}H_{27}}_{\diagdown C_{13}H_{27}}$	Tri-tridecylamine, branched	564
Alamine 336	$C_8H_{17}-M{\diagup}^{C_8H_{17}}_{\diagdown C_8H_{17}}$	Tri-C_8–C_{10}-amine, straight chain	392
Aliquat 366	$\left[\,C_8H_{17}-N{\diagup}^{C_8H_{17}}_{\diagdown C_8H_{17}}\,\Big|\,CH_3\,\right]$	Quaternary ammonium chloride	442
TBP	$\begin{array}{c} C_4M_9-O \\ C_4M_9-O \end{array}-\!\!\!\underset{O}{\overset{O}{P}}\!\!\!-O-C_4M_9$	Tributylphosphate	266

M2EMP	$C_4H_9 - CM - CH_2 - O - P = O - OM$ (with C_2H_5 and $C_4H_9 - CH - CH_2 - O$, C_2H_5) Di-(2-ethylhexyl)phosphoric acid	322
Versatic 10	C_9M_{19} — CH, O C_{10} carboxylic acid, branched	172
NA-SUL AS-50	$\left[\right] SO_3^-$ Ammonium dinonylnaphthalene sulfonate (50%) in kerosene (C_9M_{19}, C_9M_{19}, N)	462
2-ethylhexanol	$CH - CH - C_4H_8 - CH$ (C_2H_5) 2-ethylhexanol	130
LIX 64N	$C_4H_9 - CH - C = CH - CH - C_4H_9$ (MCH, CH, C_2H_5, C_2H_5) LIX 63	
	Substituted axines (50%) in kerosene	273–339
	LIX 6SN (C_9H_{19}, OH, C)	

Source: From Ref. 16.

Table 2 Biocompatibility of Untreated Secondary and Tertiary Amines with *Lactobacillus delbreuckii*

Solvent	Solvent concentration	% Glucose conversion	$Y_{x/s}$	$Y_{p/s}$	μ_{max} (h^{-1})
Control (average)	—	100	0.14	1.05	0.58
Adogen 283	10%	100	0.14	1.1	0.55
(ditridecylamine)	saturated	0	0	0	0
40 wt.% Adogen 283	10%	100	0.14	1.3	0.43
in kerosene	saturated	0	0	0	0
Adogen 364	10%	100	0.12	1.1	0.45
[tri-(C_8–c_{10})-amine)]	saturated	5	0	0	0
Trioctylamine	10%	100	0.11	1.1	0.51
	saturated	0	0	0	0
Adogen 381	10%	100	0.12	0.95	0.62
(triisooctylamine)	saturated	100	0.12	0.9	0.51
Adogen 363	10%	100	0.16	1.2	0.60
(trilaural amine)	saturated	100	0.13	1.1	0.31
40 wt.% Adogen 364	10%	100	0.10	1.1	0.49
in 2-octanone	saturated	0	0	0	0

Source: From Ref. 16.

In general, the LIX extractants had little effect on bacterial activity but the amine extractants completely inhibited cell growth at saturation concentrations.

Playne and Smith (18) investigated the effects of 30 organics on a commercial inoculum of facultatively anaerobic, acid-producing bacteria. At saturation concentrations, no inhibition was observed for *n*-hexane, isooctane, *n*-decane, kerosene, dibutylphthalate, diisoamylphthalate, tributylphosphate, tritolylphosphate, TOPO, Freon 113, Aliquat 336, diiosoamyl ether, or trioctylamine. Primene JMT and Amberlite LA2 were partially toxic, and isoamyl alcohol, hexanol, octanol, 2-ethylhexanol, dodecanol, diethyl ketone, dipropyl ketone, MIBK, benzyne toluene, *o*-xylene, nitrobenzyne, and isoamylacetate were toxic to the cells.

These studies suggest extraction solvents that are likely to be biocompatible with the cells of a potential extractive fermentation process. However, solvent biocompatibility depends on the organism used in the process. Widely varying results were found depending on whether yeast or bacteria were used and on the particular strain of bacteria. Therefore, potential extraction solvents must be tested with the process microorganism before biocompatibility can be assured.

2. Solvent Capacity and Selectivity

In addition to satisfying biocompatibility, the extraction solvent should also have a high capacity for the inhibitory fermentation products. Extraction capacity is normally given by the distribution coefficient m, defined as

$$m = \frac{\text{concentration of solute in the organic phase}}{\text{concentration of solute in the aqueous phase}} \qquad (1)$$

For a "physical" solvent, such as a hydrocarbon, the distribution coefficient is usually a constant for a given solute and solvent over an appreciable range of solute concentration. For specifically interacting "chemical" solvents, such as phosphorus-based extractants (TOPO and tributylphosphate) or amines, the distribution coefficient can be a strong function of solute concentration (16). It is desirable to use a solvent that has a distribution coefficient greater than unity to minimize the amount of solvent employed and the size of extraction equipment (19).

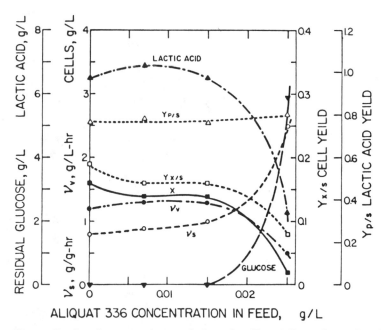

Figure 2 Continuous culture of *Lactobacillus delbreuckii* in broth containing Aliquat 336 (From Ref. 16.)

Table 3 Solvents Tested for Toxicity to *Thiobacillus ferrooxidans*

LIX 63

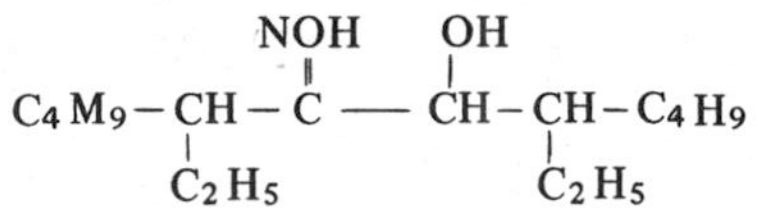

$$C_4M_9-CH-C-CH-CH-C_4H_9$$

with NOH, OH, and C_2H_5 substituents

LIX 65N

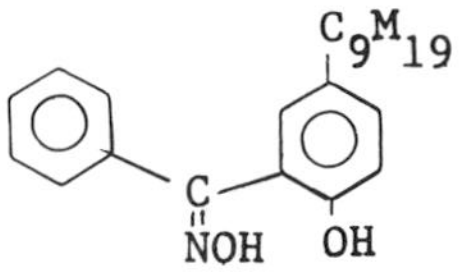

LIX 70

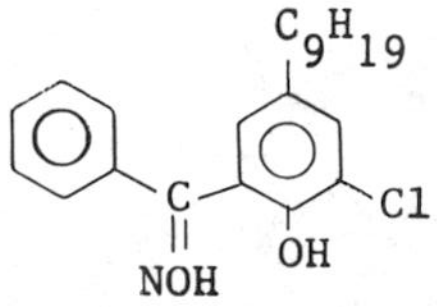

$+ \; LIX \; 63$

LIX 71

LIX 70 + LIX 65N

LIX 73

LIX 70 + LIX 65N + LIX 63

KELEX 100

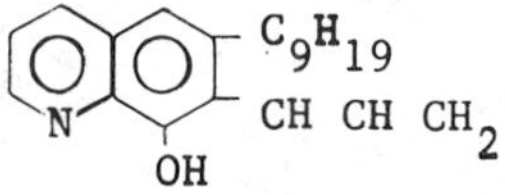

Nonylphenol

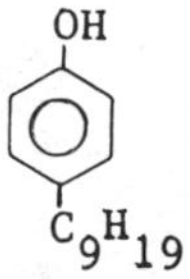

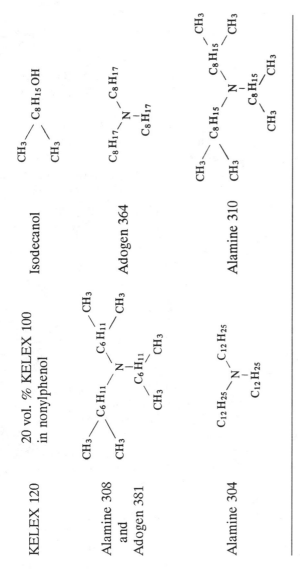

Source: From Ref. 16

Unfortunately, the solvents with the highest distribution coefficients also tend to be the most toxic to fermenting microorganisms. For example, in the extractive fermentation of lactic acid produced by *L. delbreuckii*, the best extractants were found to be tertiary and quaternary amines (16). These solvents, however, are highly toxic to the cells. Nontoxic solvents, on the other hand, have very low distribution coefficients for lactic acid (16). To provide a high distribution coefficient, a solvent must either physically resemble the solute or interact specifically with the solute via such forces as hydrogen bonding or Lewis acid-base interactions. If the solvent physically resembles the inhibitory products, however, it is likely to be inhibitory to the cells; on the other hand, if the solvent can interact specifically with the products, it probably interacts with the cell membrane and surface-bound proteins. In most extractive fermentation processes, therefore, a compromise must be made between solvent biocompatibility and extraction capacity.

It is possible to increase the distribution coefficients of weakly interacting nontoxic solvents by adding inorganic salts to the broth (20,31). Addition of salts to the broth "salts out" organic products into the extraction solvent. For dilute solvents, the increase in the distribution coefficients due to salts can be estimated from Eq. (2):

$$K_D^s = \left(K_D + \frac{18V_b}{M_oV_o} \right) e^{(2.303k_sC_a)} - \frac{18V_b}{M_oV_o} \tag{2}$$

where K_D^s is the distribution coefficient (g solute/kg solvent)/(g solute/kg broth) when salts are in the broth, K_D is the distribution coefficient without salts, V_b and V_o are the volume of the aqueous phase and that of the organic phase, respectively, M_o is the molecular weight of the organic solvent, k_s is Setchenow's constant, and C_a is the concentration of salt in the aqueous phase in gmol/L. Setchenow's constant k_s depends on the particular salt dissolved in the aqueous phase. For example, for the distribution of *n*-hexanol between water and organic solvents, k_s varies from 0.166 for NaBr to 0.568 for $MgSO_4$ (22). The presence of salts in the broth can significantly alter the distribution coefficients of organic solvents. For example, when $K_D = 1$, $V_b = V_o$, $M_o = 200$, $k_s = 0.3$, and $C_a = 1$ gmol/L, the distribution coefficient is doubled to 2.09.

Unfortunately, most bacteria can tolerate only a narrow range of salt concentrations (23). More important, many ions specifically inhibit enzyme activity and interfere with cellular transport processes (24) so that the addition of salts to the broth may inhibit cell growth and function. Also, some salts "salt in" organic products.

Selectivity of the extraction solvent for the fermentation products is also an important consideration in selecting an appropriate solvent. The selectivity of an extraction solvent for solute i over solute j, S_{ij}, is defined as

$$S_{ij} = \frac{K_{Di}}{K_{Dj}} \tag{3}$$

where K_{Di} and K_{Dj} are the distribution coefficients of solutes i and j. Compound j is often water in extractive fermentation. For example, in the extractive fermentation of ethanol, it is desirable to extract as little water as possible so that pure ethanol can be distilled from the solvent without having to break the ethanol-water azeotrope. In other fermentations, it is desirable to minimize the extraction of by-products. For example, in the extractive fermentation of acetone and butanol, it is advantageous to prevent the extraction of acetic and butyric acids so that these intermediates can be converted to products, thereby simplifying downstream processing. The solvent should not extract the substrate or nutrients from the broth. This is especially important when the solvent is regenerated by distillation because toxic compounds can be formed when sugars are heated to high temperatures.

Unfortunately, many extraction solvents that have high distribution coefficients have poor selectivities (20) and a compromise must usually be reached between solvent capacity and selectivity. It is difficult to predict accurately the distribution coefficient and selectivity of solvents, but some initial screening can be accomplished using group-contribution models, such as UNIFAC (20). Accurate values of selectivity and capacity, however, can only be obtaind by experiment.

3. Solvent Physical Properties

Physical properties of the solvent are important during the extraction step of the extractive fermentation process and in later processing steps. The most important properties to consider are solvent viscosity, interfacial tension, volatility, water solubility, and stability.

Solvent viscosity is important because the rate of mass transfer in the solvent is strongly dependent on viscosity (25). In addition, separation of the broth and solvent after extraction may be difficult if the solvent is highly viscous. It is possible to dilute highly viscous solvents in low-viscosity diluents, such as kerosene, to give a solvent mixture of intermediate viscosity.

Interfacial tension is also important in extraction processes. If interfacial tension is very high, large amounts of energy must be expended to disperse the solvent in the broth. Extremely low interfacial tensions are also undesirable because long phase-disengagement times are then required.

The volatility of the extraction solvent must also be considered. If the fermentation products are to be recovered from the extraction solvent by distillation, a choice must be made between a solvent that is more or less volatile than the extracted products. Because most fermentation products are produced in dilute solution, it is often uneconomical to distill large volumes of solvent from the products. On the other hand, if the products are volatile and heat stable, it may be more economical to use a high-boiling solvent so that only the products are distilled overhead.

Water solubility of the extraction solvent should be low to prevent solvent losses. Although volatile solvents can be recovered from the broth by steam stripping, the recovery of nonvolatile solvents is difficult and usually uneconomical.

Finally, the solvent should be stable and unreactive with all the components of the fermentation broth. In addition, the solvent should not degrade or react during solvent regeneration. If distillation is used, the solvent must also be heat-stable. Regardless of how the solvent is regenerated, it should be noncorrosive.

4. Supercritical Fluids as Extractants

Most solvents of interest for extractive bioconversions are liquids at the conditions in which the extraction takes place (usually near room temperature and ambient pressure). However, it may also be advantageous to consider another potentially interesting class of solvents: supercritical fluids. A supercritical fluid (also called a dense gas) is a fluid at a temperature and pressure somewhat above its critical temperature and pressure.

Supercritical fluids exhibit several properties that make them attractive for extraction. Although their densities may approach the densities of liquid solvents, the diffusivities of solutes in supercritical solvents may be one to two orders of magnitude higher than those in liquid solvents. In addition, supercritical fluids have viscosities about one order of magnitude lower than those found in liquids. Relatively high diffusivity and relatively low viscosity lead to high rates of mass transfer, resulting in more efficient extraction processes.

However, the most interesting property of a supercritical solvent is the high sensitivity of solubility to small changes in pressure or temperature. In the region near the critical point, small changes in the supercritical fluid's temperature and/or pressure can sometimes result in 100-fold changes in solubility. Solvent regeneration and product removal are thus simplified and may be achieved by using either a modest temperature or pressure swing, or both. Costly distillation steps may be avoided, and if the product is a solid, solvent loss by entrainment may be greatly reduced.

When choosing a supercritical solvent, the criteria are the same as those for liquid solvents: solvent biocompatibility, cost, and distribution coefficients must be examined. In addition, it is desirable to choose a fluid with a critical temperature slightly above room temperature, to avoid either refrigeration costs or thermal damage to labile products. Table 4 lists several fluids with critical temperatures near room temperature.

To date, the supercritical solvent that has attracted the most attention is supercritical carbon dioxide. Carbon dioxide offers several advantages for

Table 4 Potential Supercritical Solvents for Extractive Bioconversions

Solvent	Critical temperature ($^\circ$C)	Critical pressure (atm)
Carbon dioxide	31.1	73
Ethane	32.3	48
Ethylene	9.5	49
Nitrous oxide	36.5	70
Trifluoromethane	25.9	46

biochemical extraction processes: the critical temperature is 31.1°C, allowing thermally labile materials to be processed without thermal denaturation or decomposition. It is nonflammable and inexpensive and has a low toxicity. Carbon dioxide is inert, making unwanted side reactions with the solvent unlikely.

Although supercritical carbon dioxide is currently used as an extractant for biochemicals (e.g., caffeine from coffee and oils from fish), little information is available about the suitability of supercritical solvents for extractive bioconversions. At least one study on the toxicity of supercritical fluids to whole cells is currently underway, but little as yet is known about the effect of long exposure to supercritical fluids on cell growth, metabolism, or viability. However, it has been shown by several authors [Randolph et al. (26), Hammond et al. (27), and Nakamura and Yano (28)] that enzymes may be active in moist supercritical carbon dioxide. Cholesterol oxidase from *Gloeocysticum chrysocreas* is stable and active for at least 3 days in carbon dioxide at 100 atm and 35°C. Other enzymes that have shown activity in supercritical carbon dioxide include alkaline phosphatase (26), polyphenyl oxidase (27), lipase from *Rhizopus arrhizus* (29), and cholesterol oxidases from *Streptomyces, Pseudomonas,* and *Norcardia* (30).

C. Downstream Processing

When selecting an appropriate solvent for use in extractive fermentation, some thought must be given to subsequent downstream processing steps. There are two common methods for regenerating the solvent: distillation and back-extraction. If distillation is used, the choice must be made between using a high- and a low-boiling solvent. Figure 3 shows simplified flow sheets for processes using low-boiling and high-boiling extraction solvents. If a low-boiling solvent is used, it is distilled from the products, condensed, and recycled to the fermentor or extraction column. If a high-boiling solvent is used, the products are vaporized and recovered as an overhead product while the solvent from the column bottoms is recycled to the fermentor.

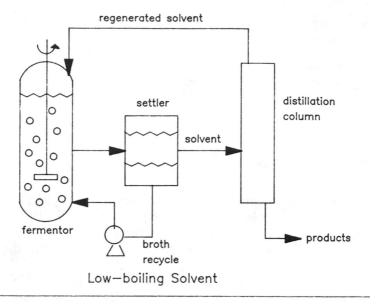

Low—boiling Solvent

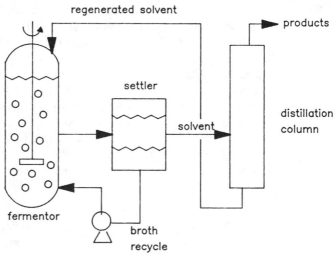

High—boiling Solvent

Figure 3 Schematic diagram of solvent regeneration using a low- or high-boiling solvent.

The advantage of using a low-boiling solvent is that relatively low temperatures can be maintained in the distillation column, especially if vacuum operation is used. This is important if the fermentation products are heat sensitive. In addition, the solvent is continuously purified in the distillation step so that toxic compounds are reduced in the solvent returned to the fermentor.

The primary disadvantages of using low-boiling solvents are that (1) they often have appreciable water solubilities and thus may be toxic to the cells, and (2) considerable energy must be expended to vaporize the large volumes of solvent from the dilute products. Much less energy is required to distill dilute products from a high-boiling solvent. High-boiling solvents are less water soluble, and toxicity is therefore reduced. The use of a high-boiling solvent, however, requires that the fermention products be heat stable. In addition, nonvolatile toxins may accumulate in the solvent, requiring periodic replacement or further regeneration of the solvent.

If the solvent is essentially nonvolatile or if the fermentation products are heat sensitive, it may be preferable to regenerate the solvent by back-extraction. Figure 4 shows a schematic diagram of an extractive fermentation process using back-extraction to regenerate the extraction solvent. Normally the products are back-extracted into water or a basic solution, such as aqueous ammonium hydroxide. If water is used to regenerate the solvent, it may be beneficial to use hot water in the back-extraction because most products are more soluble in hot

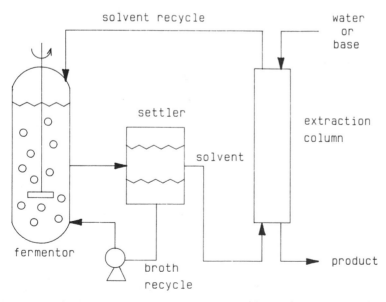

Figure 4   Schematic diagram of solvent regeneration by back-extraction.

water and the final product concentration is thus increased. Water can be subsequently removed by evaporation, under vacuum if necessary. If the fermentation products are organic acids, such as citric or gluconic acids, concentrated products can be obtained by back-extraction into an alkaline solution. An organic salt product can be obtained directly by evaporation and the organic acid product can be produced by acidifying the salt.

All these factors (solvent biocompatibility, extraction capacity and selectivity, the physical properties of the solvent, and downstream processing) must be carefully examined during the solvent selection process. Although many of the decisions must rely on engineering judgment and experience, solvent selection can be facilitated by modeling the extractive fermentation process. The next section discusses the design and modeling of extractive fermentation processes.

D. Mathematical Models for Extractive Fermentation

Kollerup and Daugulis (31) have recently modeled extractive fermentation in which the extraction solvent is added directly to a continuous fermentation. Figure 5 shows a schematic diagram of the extractive fermentation process

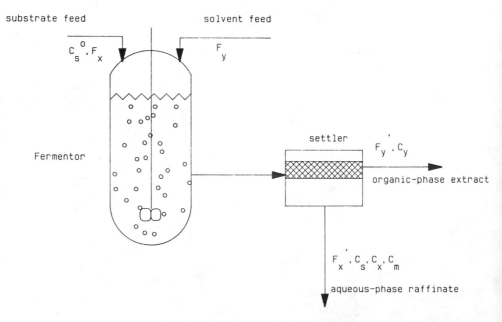

Figure 5 Schematic diagram of extractive fermentation employing direct solvent addition. Concentrations and flow rates are used in eqs. (4) through (13).

considered here. Solvent and nutrients are introduced into the fermentor at flow rates F_y and F_x. Inhibitory products formed during fermentation are extracted into drops of organic solvent formed by rapidly agitating the contents of the fermentor. Broth and solvent are continuously removed from the fermentor and separated into two phases. The flow rates of solvent and broth from the settler are represented by F_y' and F_x', respectively. Aqueous- and organic-phase flow rates are expressed as dilution rates based on the volume V of the aqeuous phase in the fermentor:

$$D_x = \frac{F_x}{V} \tag{4}$$

$$D_x' = \frac{F_x'}{V} \tag{5}$$

$$D_y = \frac{F_y}{V} \tag{6}$$

The concentration C_x of product in the aqueous-phase effluent from the fermentor is given by

$$G_1 C_x^2 + G_2 C_x + G_3 = 0 \tag{7}$$

where

$$G_1 = K_c' \left[\frac{\mu_m K_x}{\rho_p} \frac{C_s}{K_s + C_s} - D_y + \frac{g'}{Y_{p/s} C_s^0} \left(D_y - \frac{\mu_m K_x}{\rho_p} \frac{C_s}{K_s + C_s} \right) \right] \tag{8}$$

$$G_2 = K_x \left[\frac{\mu_m C_s}{K_s + C_s} \left(\frac{K_c'}{\rho_p} g - \tau_2 \right) - K_c' D_y + \frac{g'}{Y_{p/s} C_s^0} \right.$$
$$\left. \left(K_c' D_y - \mu_m \frac{C_s}{K_s + C_s} \frac{K_c' Y_{p/s} C_s}{\rho_p} - 1 \right) \right] \tag{9}$$

$$G_3 = \mu_m K_x \frac{C_s}{K_s + C_s} \left(\frac{C_s}{C_s^0} g' - g \right) \tag{10}$$

$$g = 1000 + C_s(\tau_1 - Y_{c/s}) \tag{11}$$

$$g' = 1000 + C_s^0(\tau_1 - Y_{c/s}) \tag{12}$$

where K_c' is the effective distribution coefficient in the fermentor given by $K_c' = \eta m$, where m is the distribution coefficient defined in Eq. (1) and η is the degree of equilibrium mass transfer of product between the two phases; τ_1 and τ_2 are coefficients used to estimate the density ρ of the fermentation broth.

$$\rho = 1000 + \tau_1 C_s + \tau_2 C_x \tag{13}$$

By specifying the flow rates of solvent and substrate into the fermentor and the concentration of substrate in the feed and effluent of the fermentor, the product concentration in the aqueous-phase effluent can be calculated from Eq. (7).

Kollerup and Daugulis (31) have used these equations to estimate the effect of different process variables on the productivity of an extractive fermentation process to produce ethanol. Figure 6 shows the effect of the ethanol distribution coefficient of the extraction solvent on overall ethanol productivity in the continuous extractive fermentation of 250 g/L glucose feed. Table 5 gives the parameters used to obtain the curves shown in Fig. 6.

It is clear from Fig. 6 that the solvent distribution coefficient strongly affects the productivity of the fermentation. For example, raising the solvent distribution coefficient from 0.1 to 1.0 increases the maximum ethanol productivity from 16.8 to 33.3 g/L·h. The results in Table 5, however, also point out some of the problems of carrying out extractive fermentation with direct solvent addition to the fermentor. Even for a solvent with an infinite distribution coefficient, the flow rate of solvent into the fermentor is several times that of the aqueous feed. In addition, the concentration of ethanol in the solvent C_y is several times lower than that obtained in a regular batch or continuous culture. Also, for solvents with finite distribution coefficients, since a large fraction of the product remains in the broth leaving the settler (C_x), that broth must be further processed by other means.

A model is also available for extractive fermentation using an external extraction vessel for broth and solvent (32). In the design of this type of extractive fermentation process, the primary complication is that fermentation products continue to be formed in the extraction column. The behavior of differential contactors in extractive fermentation, however, can be estimated for two

Table 5 Parameters Used for Gi. 6 at Maximum Volumetric Productivity[a]

m	D_x (h^{-1})	D_y (h^{-1})	C^0 (g/L)	C (g/L)	C_x g/L)	C_y (g/L)	Conversion (%)	Volumetric productivity (g/L·h)
0.0	0.082	0.0	250	33.0	98.6	0.0	86.6	8.15
0.1	0.153	3.0	250	11.0	48.9	3.5	95.9	16.8
0.5	0.256	3.0	250	15.0	19.2	7.7	94.5	27.7
1.0	0.304	3.0	250	14.0	12.3	9.9	95.0	33.3
∞	0.478	3.0	250	14.6	0.0	16.9	94.8	52.0

[a]Fixed parameters: $\tau_1 = 0.411$; $\tau_2 = 0.162$; $Y_{x/c} = 0.10$; $Y_{p/c} = -.46$; $Y_{c/c} = 0.44$; $K_p = 23.0$ g/L; $K_c = 1.0$ g/L; $\mu_m = 0.44$ h^{-1}; $p_p = 790.7$ g/L; $\eta = 0.80$.

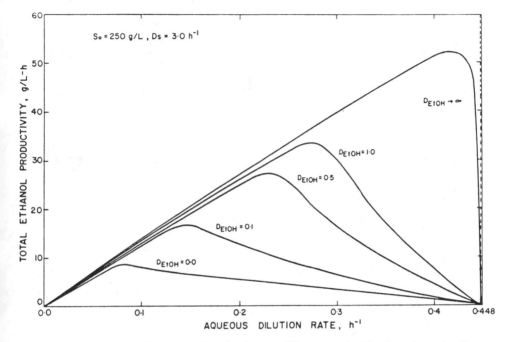

Figure 6 Effect of ethanol distribution coefficient on total ethanol productivity. (From Ref. 31.)

limiting product formation kinetics (32). When substrate concentration in the broth is high compared to the Monod constant, the rate of product formation is often independent of substrate concentration. In this case, the dimensionless concentration of product in the aqueous and organic phases leaving the extraction column is available (32).

These models are applicable when product formation is zero order in substrate concentration. When substrate concentration is low, however, product formation kinetics are often first order in substrate concentration. For first-order product formation kinetics, the dimensionless concentration profile of substrate in the extraction column is also given in Ref. 32. Although the solutions given by Roffler et al. (32) are approximations, they approach the exact solutions to within 5% for most practical cases of extractive fermentation.

The production of fermentation products by viable cells in the broth passing through the extraction column can dramatically alter the performance of the column. Figure 7 shows how the degree of extraction obtained in columns of different length depends on the volumetric rate of product formation inside the extractor for the operating conditions listed in Ref. 32. When there are no cells

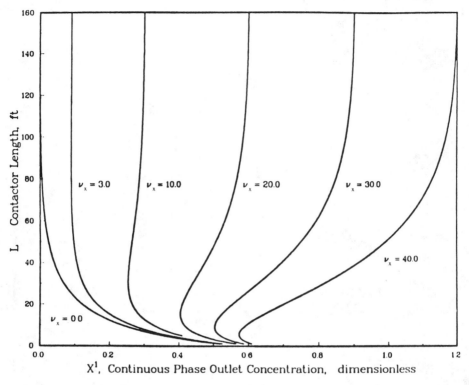

Figure 7 Contactor length required to obtain a given raffinate concentration at different rates of product formation inside the contactor. (From Ref. 16.)

in the broth ($N_X = 0$), the concentration of product in the aqueous-phase efflu-ent (X^1) can be reduced to any concentration by increasing the contactor length. When cells are present in the aqueous phase, however, there is a limit to the degree of extraction that can be obtained. For example, the minimum dimensionless outlet concentration in the aqueous phase is about 0.08 when $\nu_X = 3$ g/L·h and 0.24 when $\nu_X = 10$ g/L·h. In addition, the greatest degree of extrac-tion occurs for finite column lengths when there is product formation inside the column. This maximum follows from the counteracting effects of continuous extraction of products from the broth by the solvent and excretion of products into the broth by the cells. Whenever the rate of product formation exceeds the rate of product extraction, the concentration of product in the aqueous-phase effluent increases. It is therefore important to consider the effects of product formation on extractor performance and design in extractive fermentation.

E. Practice of Extractive Fermentation

Extractive fermentation has only been demonstrated in laboratory-scale experiments. It has been primarily applied to the production of bulk chemicals, such as ethanol and butanol. In addition, most experimental extractive fermentation systems have been operated with direct solvent addition to the fermentor.

In one of the first studies of extractive fermentation, Finn (5) tried to relieve product inhibition of prodigiosin on a strain of *Serratia* by extraction into kerosene; however, production was not increased. More success was achieved in studies when ethanol produced by *S. cerevisiae* was extracted into dodecanol (12,33) or dibutylphthalate (7). Ribaud (34) showed that ethanol productivity could be slightly increased by extracting ethanol in situ with dibutylphthalate. The ethanol distribution coefficient in dibutylphthalate, however, is only 0.13; large volumes of solvent would be required in an industrial process. Minier and Goma (12,33) used dodecanol to continuously extract ethanol produced by yeast cells immobilized on pieces of brick in a tower fermentor. Using extractive fermentation, a feed containing 40% wt./wt. glucose was completely fermented at a volumetric productivity five times greater than the control. Dodecanol, unfortunately, has a low capacity for ethanol (m = 0.35) and the solvent flow rate was 17 times that of the sugar feed into the fermentor.

Extractive fermentation has also been used to extract inhibitory organic acids, including butyric, valeric, and caproic acids, from a suppressed-methane anaerobic fermentation (35). The extraction solvent, kerosene, was regenerated by back-extraction into a basic solution. Concentration of the salt of the desired acid in the basic solution was 10 times greater than in the fermentor. Inhibitory acetic and propanoic acids, however, are almost insoluble in kerosene and therefore accumulate in the broth to toxic levels.

The greatest amount of research on extractive fermentation has been conducted on the production of acetone and butanol by *C. acetobutylicum*. Growth of *C. acetobutylicum* is extremely sensitive to the accumulation of butanol in the broth; the cells are totally inhibited by butanol concentrations of only 10–15 g/L (1,2). Hashimoto (7) reported slight increases in fermentation productivity when butanol was extracted into dibutylphthalate during fermentation. Traxler et al. (9) extracted butanol with hexanol, octanol, dodecanol, hexadecanol, ethylcaproate, cyclohexanol, or 4-methylcyclohexanol during fermentation. Slight increases in butanol yield were found when hexadecanol or ethylcaproate were used, but yield decreased with the other solvents.

Ishii et al. (11) and Taya et al. (36) used extractive fermentation to relieve butanol inhibition during fermentation of glucose by *C. acetobutylicum* in batch and fed-batch culture (Fig. 8a and b). Best results were obtained when oleyl alcohol (C_{16}–C_{18} unsaturated alcohol) or Guerbet alcohol (C_{20} branched-chain

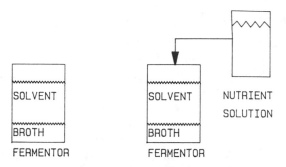

a. BATCH EXTRACTIVE FERMENTATION   b. FED–BATCH EXTRACTIVE FERMENTATION

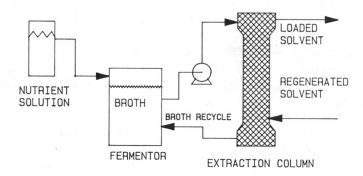

c. CONTINUOUS PRODUCT EXTRACTION FROM A FED–BATCH FERMENTATION

Figure 8 Experimental extractive fermentation apparatus.

alcohol) was used as extractant. In batch extractive fermentation, *C. acetobutylicum* was able to ferment 73 g/L glucose producing 12.9 g/L butanol compared to 29 g/L glucose and 5 g/L butanol in a regular batch culture. Because the cells were inhibited by concentrations of glucose over 80–100 g/L, Taya et al. (36) used fed-batch extractive fermentation (Fig. 8b). In a fed-batch culture, a concentrated solution of nutrients is fed to the fermentor at a controlled rate so that substrate concentration in the broth is maintained below inhibitory levels. Cells of *C. acetobutylicum* were able to ferment a feed solution containing 500 g/L in fed-batch extractive fermentation when oleyl alcohol was used as the extractant. By periodically withdrawing and adding new extraction solvent to the fermentor, butanol concentration in the broth was maintained below 2 g/L. A total of 120 g/L glucose was consumed and 20.4 g/L butanol was produced, based on final broth volume. In both batch and fed-batch extractive fermenta-tions, however, the productivity of the cells was not increased by the in situ removal of butanol.

Roffler et al. (10,37,38) have also used extractive femernation to reduce butanol inhibition during the fermentation of glucose by *C. acetobutylicum*. In batch extractive fermentation, best results were obtained when oleyl alcohol or a mixture of oleyl alcohol and benzyl benzoate were used as extraction solvents. In culture, *C. acetobutylicum* (ATCC 824) was able to ferment 82 g/L glucose and produce 15 g/L butanol in 24 h. However, 18 g/L glucose remained at the end of culture, because of the inhibition of the cells by butanol. When 1.1 L of oleyl alcohol was added to 2 L of broth during h 8 of fermentation, the aqueous-phase butanol concentration was maintained below 7 g/L and glucose conversion was rapid and complete; 98 g/L of glucose were consumed in less than 20 h. Butanol productivity increased from 0.58 g/L·h in a regular batch culture to 0.72 g/L·h in batch extractive fermentation using oleyl alcohol as the extraction solvent.

Roffler et al. (37) also demonstrated production of acetone and butanol in fed-batch extractive fermentation (Fig. 8b). Figure 9 shows the results of a fed-batch extractive fermentation in which 2.25 L of oleyl alcohol was added to 1.5 L of broth. A concentrated nutrient solution containing 339 g/L glucose was metered into the fermentor at a rate such that the glucose concentration in the broth was maintained at 10–20 g/L throughout the fermentation. At the end of fermentation, the aqueous phase contained 8.8 g/L butanol and 10.4 g/L ace-tone and the organic phase contained 35 g/L butanol and 8.2 g/L acetone. A total of 207 g/L glucose was consumed and 45 g/L butanol produced (based on final broth volume), representing a 300% increase over batch culture. Nutrient solutions containing up to 500 g/L glucose could be fermented at productivities as high as 1.5 g/L·h, 150% greater than those in batch culture.

Extractive fermentation of acetone and butanol has also been carried out using an external extraction vessel (38). Figure 8c shows a schematic diagram of the experimental apparatus used. Concentrated nutrient solution, containing 300 g/L glucose, was metered into the fermentor on demand. Broth containing viable cells of *C. acetobutylicum* (ATCC 824) was metered into the top of a Karr recip-rocating-plate extraction column while the extraction solvent, oleyl alcohol, was metered into the bottom of the column. Inhibitory products in the fermentation broth were extracted into solvent rising in the extraction column. Extracted broth was returned to the fermentor while the solvent leaving the column was regenerated by steam stripping and returned to the extraction column for reuse. Figure 10 shows results of the extractive fermentation of acetone and butanol carried out in this apparatus. The top section of Fig. 10 shows the concentration of acetone and butanol in the oleyl aclohol leaving the extraction column. The organic-phase butanol concentration averaged around 12 g/L. The concentration

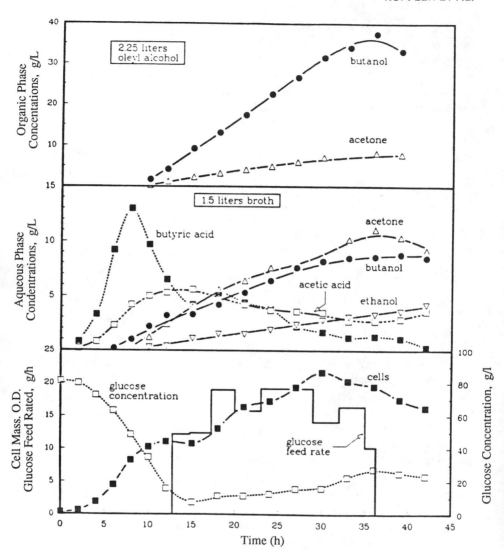

Figure 9 Fed-batch extractive fermentation of *Clostridium acetobutylicum* using 2.25 L oleyl alcohol. Initial broth volume was 1.5 L. Glucose concentration in the feed was 339 g/L. (From Ref. 16.)

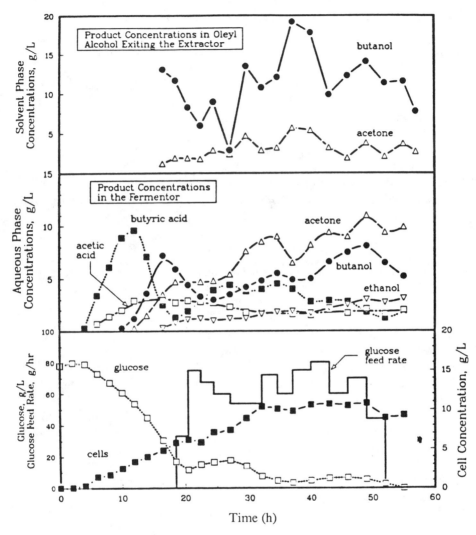

Figure 10 Product and substrate concentrations during the continuous extraction of acetone and butanol from a batch-fed culture of *Clostridium acetobutylicum*. Oleyl alcohol was used as the extraction solvent. The concentrated feed solution contained 300 g/L of glucose. (From Ref. 16.)

of butanol in the broth was maintained below 8 g/L throughout fermentation, allowing the fermentation to proceed for over 50 h at a volumetric productivity about twice that of regular batch culture; 214 g/L of glucose were consumed and 40 g/L of butanol was produced, based on the final broth volume.

Extractive fermentation has rarely been used for products other than low-molecular-weight alcohols and organic acids. Becker et al. (39), however, have shown that extractive fermentation can benefit the fermentation of more complex products. They applied extractive fermentation to plant tissue cultures of *Matricaria chamomilla, Pimpinella anisum,* and *Valeriana wallichii.* These cultures produce essential oils and other lipophilic products that normally degrade over the course of fermentation. By extracting these products into a C_8-C_{10} triglyceride extraction solvent, product degradation was reduced and final product yields increased.

F. Extractive Fermentation Economics

Little information is available on the economic potential of extractive fermentation. A study has recently been completed, however, comparing the economics of producing butanol from molasses by conventional batch fermentation and by extractive fermentation (40). Figure 11 shows a schematic diagram of a conventional fermentation process to produce butanol. Molasses containing 55 wt.% fermentable sugars is diluted to 60 g/L sugar and mixed with nutrients before being continuously sterilized and fed into batch fermentors.The fermentors are inoculated with actively growing cells of *C. acetobutylicum* produced in prefermentors. After 30 h of fermentation, the broth, containing 13.7 g/L butanol, 5.4 g/L acetone, 1.5 g/L ethanol, 3 g/L cells, and 0.5 g/L acetic and butyric acids, is discharged into the broth surge tank. Butanol, acetone, and ethanol are first stripped from the broth and then separated in a series of distillation columns. The stripped broth, containing acetic and butyric acids, cells, proteins, and nonfermentable molasses solids, is evaporated to 50 wt.% solids in multiple-effect evaporators and then dried to 85 wt.% solids in a rotary drier to give a dried stillage product that can be used as an animal-feed supplement.

A batch plant capable of producing 200 million pounds of butanol per year was estimated to require a capital investment of $154 million (40). Table 6 shows a breakdown of the manufacturing costs for the batch fermentation process. The rational price for butanol for an 18% discounted cash flow rate of return on investment was estimated to be 61.7 cents/pound.

Figure 12 shows a schematic diagram of an extractive fermentation process to produce 200×10^6 pounds of butanol annually. Fed-batch operation of the fermentors is used to prevent substrate inhibition of the cells by high concentrations of sugars in the fermentor. Molasses diluted to 500 g/L sugar is fed to the fermentors as needed to maintain residual sugar at 12-15 g/L. A concentrated

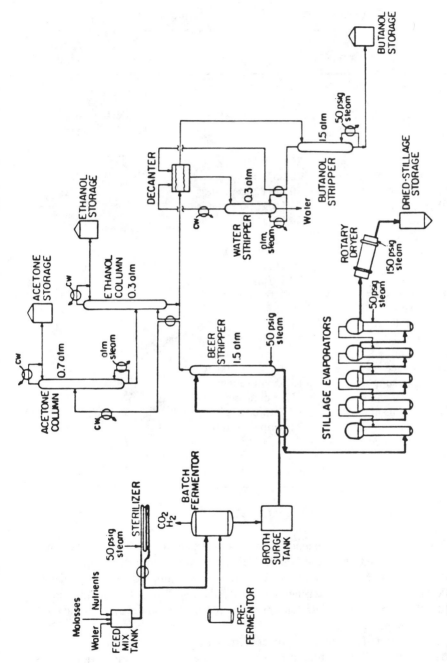

Figure 11 Schematic diagram of the acetone-butanol batch fermentation process. (From Ref. 16.)

Table 6   Manufacturing Costs for Batch Fermentation of Butanol

Item	Basis	Cents/pound butanol
Raw materials		
Nutrients	1 /lb BuOH	1.0
Water	$1.10/1000 gallons	0.8
Molasses	$100/ton, 55% sugar	41.2
Utilities		
Power	$0.08/kWh	0.6
Cooling water	$0.25/1000 gallons	0.4
Atm steam	$0.85/1000 pounds	0.2
50 psig steam	$3.50/1000 pounds	11.0
150 psig steam	$4.00/1000 pounds	0.8
Coproducts		
Acetone	$0.27/pound	(10.6)
Ethanol	$0.26/pound	(2.8)
Hydrogen	$0.10/pound	(0.7)
Dried stillage	$60/ton, 85% solid	(9.5)
Atm steam	$0.85/1000 pounds	(1.3)
Total variable cost		
Operating labor	$12/labor h, 42 operators	2.1
Supervision	15% operating labor	0.3
Maintenance	4% fixed capital cost	2.8
Operating supplies	15% maintenance	0.4
Laboratory charges	15% operating labor	0.3
Taxes and insurance	1.5% fixed capital cost	1.0
Plant overhead	22 people, $80,000/year	0.9
Total fixed cost		7.8
Capital charges (30% of total capital cost)		22.8
Rational price (variable + fixed cost + capital charges)		61.7

Source: From Ref. 16.

feed is used to minimize the amount of water introduced into the process. Whole broth is circulated to the extraction column, where inhibitory fermentation products are extracted into an organic solvent containing 50 wt.% oleyl alcohol in decane. The solvent is regenerated by distillation. Products remaining in the broth at the end of fermentation are recovered in the same way as in the batch fermentation process.

The extractive fermentation process requires an estimated capital investiment of $125 million. The manufacturing costs of the extractive fermentation process are listed in Table 7. The rational price of butanol produced by extractive

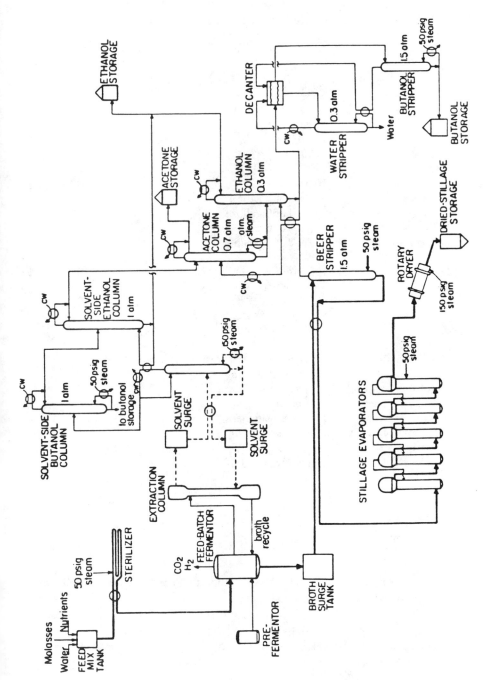

Figure 12 Schematic diagram of the acetone-butanol extractive fermentation process. (From Ref. 16.)

Table 7 Manufacturing Costs for Fed-Batch Extractive Fermentation of
Butanol

Item	Basis	Cents/pound butanol
Raw materials		
Nutrients	1 /pound butanol	1.0
Water	$1.10/1000 gallons	0.1
Molasses	$100/ton	41.2
Utilities		
Power	$0.8/kWh	0.9
Cooling water	$0.25/1000 gallons	0.3
Atm	$0.85/1000 gallons	0.1
50 psig steam	$3.50/1000 pounds	1.6
150 psig steam	$4.00/1000 pounds	1.9
Coproducts		
Acetone	$0.27/pound	(10.6)
Ethanol	$0.26/pound	(2.8)
Hydrogen	$0.10/pound	(0.7)
Dried stillage	$60/ton, 85% solids	(9.5)
Atm steam	$0.85/1000 pounds	(0.2)
Total variable cost		23.3
Operating labor	$12/labor h, 44 operators per shift	2.2
Supervision	15% operating labor	0.3
Maintenance	4% fixed capital cost	2.2
Operating supplies	15% maintenance	0.3
Laboratory charges	15% Operating labor	0.3
Solvent makeup	15% solvent inventory	0.1
Insurance and taxes	1.5% fixed capital	0.8
Plant overhead	22 people, $80,000/year	0.9
Total fixed costs		7.1
Capital charges (30% of total cost)		18.5
Rational price (variable + fixed cost + capital charges)		48.9

Source: From Ref. 16.

fermentation was estimated to be 48.9 cents/pound, 20% lower than the cost of
butanol produced by conventional batch fermentation of molasses. Major
savings in the extractive process resulted from the use of fewer fermentors due
to increased fermentor productivity and from reduced energy and capital
requirements in the stillage-handling section of the plant due to reduced water
requirements.

III. ENZYMATIC REACTIONS IN AQUEOUS-ORGANIC SOLVENT, TWO-PHASE SYSTEMS

Enzymes provide very high reaction velocities, and many are highly substrate specific, making possible the rapid conversion of a single reactant. However, as with microbial fermentations, product inhibition can occur, leading to dilute product streams. Further, many reactions of interest involve water-insoluble or poorly water-soluble substrates; these transformations cannot efficiently be carried out in homogeneous aqueous solutions because of the large reactor volumes required. Other reactions are impossible in aqueous solutions because of unfavorable reaction equilibria, such as the synthesis of ester or peptide bonds using hydrolytic enzymes. One solution to these problems is to perform the enzymatic reaction in an aqueous-organic solvent two-phase system.

In a two-phase system the enzyme is contained in the aqeuous phase and the reactants and products are carried by the organic phase. The reaction occurs at the interface between the two phases or within the aqueous phase itself, depending on the enzyme. The reactants transfer from the organic to the aqueous phase, where the enzyme catalyzes the conversion to products, and the products move back into the organic phase. Contact between enzyme and product is minimized as products are in effect "extracted" into the organic phase, reducing product inhibition. Because of the increased solubilities of reactants and products in the organic solvent, higher concentrations in the reactor are attainable, leading to faster reaction rates and simplified product recovery. Further, hydrolytic enzyme reaction equilibria can be shifted toward synthesis rather than hydrolysis of ester or peptide bonds. We turn now to a brief discussion of solvent selection and reactor design for enzymatic two-phase systems.

A. Solvent Selection

When choosing an organic solvent for the two-phase system, several important factors must be considered. To minimize solvent requirements, the solvent must have a high capacity to solubilize reactants and products. However, if the enzyme has a large K_m value, a small distribution coefficient may be desired so that more substrate is available to the enzyme.

Many organic solvents strongly inhibit or denature enzymes; therefore, one must be careful to choose a solvent that does not adversely affect the enzyme. Unfortunately, it is not possible to predict the denaturing or inhibitory effects of a given organic solvent toward an enzyme; the solvent must be tested with the enzyme. However, a solvent with a low water solubility is likely to be better than one with a high water solubility since the former contacts the enzyme to a lesser extent. Other solvent characteristics to consider are cost, volatility, viscosity, flammability, and toxicity. Butler (41) and Lilly (42) provide good discussions of solvent selection for enzymatic transformations in two-liquid-phase systems.

Cremonesi et al. (43) tested the effect of several organic solvents on the stability and activity of 20-β-hydroxysteroid dehydrogenase in a two-phase system. They found butyl acetate to be the best overall solvent in terms of substrate conversion. Lugaro et al. (44) studied the oxidation of steroid hormones by fungal laccase in a wide variety of organic solvents. They concluded that the most polar solvents lead to a greater inactivation of the enzyme, and the less polar solvents are less efficient owing to lower steroid solubility. Other enzymes that have been studied in biphasic systems include chymotrypsin, thermolysin, pepsin, papain, alkaline phosphatase, lipases, and peroxidases (45). Solvent choice often results from a compromise between enzyme stability and activity on the one hand and the solubility of substrates and products on the other.

B. Reactor Design

The presence of two phases necessitates the transfer of reactants from the organic to the aqueous phase. High interfacial areas are therefore needed to lower the mass-transfer limitations of the reaction. Moderate rates of agitation generally give the maximum transformation of reactants to products, because very high rates of agitation can lead to enzyme denaturation (46).

Activity per unit volume of reactor is an important design parameter. As discussed in the previous section, solvent choice can play an important role in determining catalyst activity and stability. The ratio of organic- to aqueous-phase volume is another important consideration. If reactants and products are both contained in the organic phase, then it is desirable to use as little water as needed to preserve enzyme activity. Zaks and Klibanov (47) report that porcine pancreatic lipase catalyzes the transesterification between tributyrin and heptanol in a medium containing less than 0.02% water, and lipases from *Mucor* sp. and *C. cylindracea* also function catalytically in solutions containing less than 1% water (48). If reactants or products are soluble in the aqueous phase, the optimum ratio of phase volumes should be determined experimentally.

The operational stability of the enzyme is also of major concern because the cost of supplying fresh enzyme can be a limiting economic factor. Fortunately, enzymes in aqueous-organic two-phase systems are relatively stable if the solvent is chosen judiciously. However, it is often desirable to increase further enzyme stability by attaching it to a solid support and thereby fixing it in its active conformation. This can be done in a variety of ways; several reviews on this subject are available (49,50).

Carrea et al. (51) investigated the effect of immobilization conditions on the activity and stability of hydroxysteroid dehydrogenases immobilized to CNBr-activated Sepharose. Immobilization in the presence of NAD gave a highly stable enzyme: β-HSDH maintained 60% of its original activity 2 months after continuous use in a water-ethyl acetate system. Immobilization also resulted in increased operational stability of lipase (52). This increase in stability is often at the expense of enzyme activity, however, and some enzymes lose almost all activity when immobilized. Again, this can only be determined empirically. Wisdom et al. (53) studied the suitability of three grades of diatomaceous earth and a controlled-pore silica as support materials for lipase catalyzing the interesterification of fats. They found that Hyflo Supercel gave the highest interesterification activity, and no interestification activity was detectable when the controlled-pore silica was used.

For the enzymatic conversion of poorly water-soluble reactants, aqueous-organic solvent, two-phase systems have several possible advantages over homogenous systems:

1. High reactant and product concentrations in the reactor allow the use of reduced reactor volumes.
2. Enzyme inhibition by substrates or products is minimized.
3. Hydrolytic enzymes can be used for the synthesis of ester or peptide bonds.

Solvent selection is critical to the success of enzymatic two-phase conversions; experimental information for the system of interest is almost always needed to make the best solvent choice. Because of their unique features, many enzymatic two-phase systems are likely to be commercialized in the future.

A special case of extractive bioconversion is when the catalyst is confined in a very small volume of water in continuous organic phase. One example of such systems is extractive bioconversions in reverse micelles (54,55).

This is an intensive research area today, but it is excluded from this volume because the emphasis is on treatment of an aqueous phase with various extractions.

IV. CONCLUSIONS

Extraction with nonaqueous solvents provides a method to increase the productivity of bioconversions. Productivities may be increased by the removal of product inhibition or by preventing product degradation. A number of solvents have potential application for extractive fermentation processes. Each must be evaluated in terms of biocompatibility, capacity, selectivity, and physical properties; the evaluation may be aided by using available models for extractive fermentations.

Extraction with nonaqueous solvents may also be applied to enzymatic reactions; the same solvent selection criteria apply as those for fermentation systems.

Enzymatic reaction productivities may be increased by using a two-phase, aqueous-organic solvent system.

V. NOMENCLATURE

C_a concentration of salt in the aqueous phase, gmol/L
C_m concentration of cells in X phase, $M\ L^{-3}$
C_s concentration of substrate in X phase, $M\ L^{-3}$
C_x concentration of solute in X phase, $M\ L^{-3}$
C_y concentration of solute in Y phase, $M\ L^{-3}$
K_s sugar inhibition constant, $M\ L^{-3}$
L   total length of contactor, L
M   molecular weight
m   defined in Eq (1)
S   generalized substrate concentration in the X (feed) phase, deminsionless
S_{ij} selectivity of a solvent for solute i over solute j, dimensionless.
k_s Setchenow's constant
K_D distribution coefficient on a weight basis, dimensionless
K_{Ds} distribution coefficient (weight basis) when salts are present in the aqueous phase, dimensionless
V   volume of organic or aqueous phase, L^3
X   generalized solute concentration in X (feed) phase, dimensionless
Y   generalized solute concentration in Y (extractant) phase, dimensionless
$Y_{p/s}$ yield of product (solute) from substrate M/M, dimensionless
$Y_{x/s}$ yield of cells from substrate M/M, dimensionless
Z   z/L, fractional length within column, dimensionless
z   length within column measured from X-phase inlet, L
F_j flow rate of the Jth phase into a continuous fermentation, $L^3\ T^{-1}$
F_j' flow rate of the jth phase out of a continuous fermentation, $L^3\ T^{-1}$
D_j dilution rate of the jth phase, T^{-1}
G_i coefficients defined in Eqs. (8) through (10)
K_x, K_p product inhibition constant, $M\ L^{-3}$
g, g'   defined by Eqs. (11) and (12)
$Y_{c/s}$ yield of gaseous products from substrate, M/M, dimensionless

Greek Letters

ϵ_x σ/y, dimensionless
μ specific growth rate of cells, T^{-1}
μ_m maximum specific growth rate, T^{-1}
o_x rate of product formation, $M\ L^{-3}\ T^{-1}$
τ_i coefficients used in Eq. (13)

ρ density of fermentation broth, g/L

η degree of equilibrium mass transfer between aqueous and organic phases, dimensionless

Superscripts

0 feed inlet end, outside column ($Z = 0$)
1 extractant inlet end, outside column ($Z = 1.0$)
* value at equilibrium with other phase

Subscripts

b aqueous or broth phase
j X or Y phase
m cells
o organic phase
s substrates
p product
x X phase (feed)
y Y phase (extractant)

REFERENCES

1. Hasting, J. H., Acetone-butyl alcohol fermentation, in: *Economic Microbiology* (A. H. Rose, ed.), Vol. 2. Academic Press, New York, pp. 31–44 (1978).
2. Walton, M. T., and Martin J. L., Production of butanol-acetone by fermentation, in: *Microbiol Technology* (J. J. Peppler and D. Perlman, eds.), Vol. 1. Academic Press, New York, pp. 187–209 (1979).
3. Mattiasson, B., and Larsson, M., Extractive bioconversions with emphasis on solvent production, *Biotechnol. Genet. Eng. Rev.* 3:137–174 (1985).
4. Karr, A. E., Design, scale-up, and applications of the reciprocating plate extraction column, *Separation Sci. Technol.* 15:877–905 (1980).
5. Finn, R. K., Inhibitory cell products: Their formation and some new methods of removal, *J. Ferment. Technol.* 44:305–310 (1966).
6. Pye, E. K., Thermophilic degradation of cellulose for production of liquid fuels, in: Second Annual Symposium on Fuels from Biomass. 11:601–608 (1978).
7. Hashimoto, Y., Optimization of an extractive fermentation process for the production of butanol. M.Sc. Thesis, University of Pennsylvania, Philadelphia (1979).
8. Annual Technical Report of the Research Association for Petroleum Alternatives Development, Chiyoda-ku, Tokyo 101, Japan, September 1985.
9. Traxler, R. W., Wood, E. M., Mayer, J., and Wilson, M. P., Jr., Extractive fermentation for the production of butanol, *Develop. Ind. Microbiol.* 26: 519–525 (1985).

10. Roffler, S. R., Blanch, H. W., and Wilke, C. R., In-situ recovery of butanol during fermentation. I. Batch extractive fermentation, *Bioprocess Eng.* 2(1):1–12 (1987).

11. Ishii, S., Taya, M., and Kobayashi, T., Production of butanol by *Clostridium acetobutylicum* in extractive fermentation system, *J. Chem. Eng., Japan* 18:125–130 (1985).

12. Minier, M., and Goma, G., Ethanol production by extractive fermentation, *Biotechnol. Bioeng.* 24:1565–1579 (1982).

13. Matsumara, M., and Markl H., Application of solvent extraction to ethanol fermentation, *Appl. Microbiol. Biotechnol.* 20:371–377 (1984).

14. Datta, R., Acidogenic fermentation of corn stover, *Biotechnol. Bioeng.* 23: 61–77 (1981).

15. Dave, G., Blanck H., and Gustaffson, K., Biological effects of solvent extraction chemicals on aquatic organisms, *J. Chem. Technol. Biotechnol.* 29:249–257 (1979).

16. Roffler, S., Extractive fermentation-lactic acid and acetone/butanol production, Ph.D. Dissertation, University of California, Berkeley (1986).

17. Torma, A. E., and Itzkovitch, I. J., Influence of organic solvents on Cholacopyrite oxidation ability of *Thiobacillus ferrooxidans, Appl. Environ. Microbiol.* 32:102–107 (1976).

18. Playne, M. J., and Smith, B. R., Toxicity of organic extraction reagents to anaerobic bacteria, *Biotechnol. Bioeng.* 25:1251–1265 (1983).

19. King, C. J., *Separation Processes*, 2nd ed. McGraw-Hill, New York (1980).

20. Murphy, T. K., Blanch, H. W., and Wilke, C. R., Water recycle in extractive fermentation, *Process Biochem.* 17:6–9 (1982).

21. Leonard, R. H., Peterson, W. H., and Johnson, M. J., Lactic acid from fermentation of sulfite waste liquor, *Ind. Eng. Chem.* 40:57–67 (1948).

22. Murphy, K., Ph.D. Thesis, University of California, Berkeley (1984).

23. Rose, A. H., The environment, in: *Chemical Microbiology*, 2nd ed. Plenum Press, New York, pp. 57–101 (1968).

24. Maiorella, B. L., Fermentative ethanol production, Ph.D. Dissertation, University of California, Berkeley, pp. 234–255 (1983).

25. Sherwood, T. K., Pigford, R. L., and Wilke, C. R., *Mass Transfer* McGraw-Hill, New York (1978).

26. Randolph, T. W., Blanch, H. W., Prausnitz, J. M., and Wilke, C. R., Enzymatic catalysis in a supercritical fluid, *Biotech. Lett.* 7(5):325–328 (1985).

27. Hammond, D. A., Karel, M., Klibanov, A., and Krukonis, V. J., Enzymatic reactions in supercritical cases, *Appl. Biochem. Biotechnol.* 11:393–400 (1985).

28. Nakamura, K., and Yano, T., Japan Kokai Tokyo Koho JP 61/21098 A2 [86/21098], Jan. 29 (1986).

29. Miller, D. A., Blanch, H. W., and Prausnitz, J. M., Enzymatic interestification of triglycerides in supercritical carbon dioxide, to appear in *Ann. NY Acad. Sci.*, Proceeding of Enzyme Engineering X, Kashikojima, Japan, Sept. 24–29, 1989.

30. Randolph, T. W., Blanch, H. W., and Prausnitz, J. M., Enzyme-catalyzed oxidation of cholesterol in supercritical carbon dioxide, *AIChE J.* 34(8): 1354–1360 (1988).

31. Kollerup, F., and Daugulis, A. J., A mathematical model for ethanol production by extractive fermentation in a continuous stirred tank fermentor, *Biotechnol. Bioeng.* 27:1335–1346 (1985).

32. Roffler, S. R., Blanch, H. W., and Wilke, C. R., Mathematical models for differential contactors in extractive fermentation, *Biotech. Bioeng.* 32: 192 (1988).

33. Minier, M., and Goma, G., Production of ethanol by coupling fermentation solvent extraction. *Biotechnol. Lett.* 3(8):405–408 (1981).

34. Ribaud, J., Feasibility study on the use of extractive fermentation to enhance ethanol production in the yeast fermentation, M.Sc. Thesis, University of Pennsylvania, Philadelphia (1980).

35. Levy, P. F., Sanderson, J. E., and Wise, D. L., Development of a process for the production of liquid fuels from biomass, *Biotechnol. Bioeng. Symp.* (11):239–248 (1981).

36. Taya, M., Ishii, S., and Kobayashi, T., Monitoring and control for extractive fermentation of *Clostridium acetobutylicum*. *J. Ferment. Technol.* 63:181–187 (1985).

37. Roffler, S. R., Blanch, H. W., and Wilke, C. R., In-situ recovery of butanol during fermentation. II. Fed-batch extractive fermentation. *Bioprocess Eng.* 2(2):83–92 (1987).

38. Roffler, S. R., Blanch, H. W., and Wilke, C. R., Extractive fermentation of acetone and butanol, *Biotechnol. Bioeng.* 31:135 (1988).

39. Becker, H., Reichling, J., Bisson, W., and Herold, S., Two-phase culture—a new method to yield lipophilic secondary products from plant suspension cultures, in: Third European Congress on Biotechnology, Vol. 1, pp. 209–213, Munchen, West Germany, September 10–14 (1984).

40. Roffler, S. R., Blanch, H. W., and Wilke, C. R., Extractive fermentation of acetone and butanol. Process design and economic evaluation, *Biotechnol. Prog.* 3(3):131 (1987).

41. Butler, L. G., Enzymes in non-aqueous solvents, *Enzyme Microb. Technol.* 1:253–259 (1979).

42. Lilly, M. D., Two-liquid-phase biocatalytic reactions, *J. Chem. Technol. Biotechnol.* 32:162–169 (1982).

43. Cremonesi, P., Carrea, G., Ferrara, L., and Antonini, E., Enzymatic preparation of 20-beta-hydroxysteroids in a two-phase system, *Biotechnol. Bioeng.* 17:1101–1108 (1975).

44. Lugaro, G., Carrea, G., Cremonesi, P., Casellato, M. M., and Antonini, E., The oxidation of steroid hormones by fungal laccase in emulsion of water and organic solvents, 159:1–6 (1973).

45. Carrea, G., Biocatalysis in water-organic solvent two-phase systems, *Trends Biotechnol.* 2(4):102–106 (1984).

46. Cremonesi, P., Carrea, G., Sportoletti, G., and Antonini, E., Enzymatic dehydrogenation of steroids by beta-hydroxysteroid dehydrogenase in a two-phase system, *Arch. Biochem. Biophys.* 159:7–10 (1973).

47. Zaks, A., and Klibanov, A. M., Enzymatic catalysis in organic media at 100°C, *Science* 224:1249–1251 (1984).
48. Zaks, A., and Klibanov, A. M., Enzyme-catalyzed processes in organic solvents, *Proc. Natl. Acad. Sci. USA* 82:3192–3196 (1985).
49. Mosbach, K., ed., Immobilized enzymes, *Methods Enzymol.* 44 (1976).
50. Wingard, L. B., Katchalski-Katzir, E., and Goldstein, L., eds., Immobilized enzyme principles, *Appl. Biochem. Bioeng.* 1 (1976).
51. Carrea, G., Colombi, G., Mazzola, G., and Cremonesi, P., Immobilized hydroxysteroid dehydrogenases for the transformation of steriods in water-organic solvent systems, *Biotechnol. Bioeng.* 21:39–48 (1979).
52. Yokozeki, K., Yamanaka, S., Takinami, K., Hirose, Y., Tanaka, A., Sonomoto, K., and Fukui, S., Application of immobilized lipase to regio-specific interesterification of triglyceride in organic solvent, *Eur. J. Appl. Microbiol. Biotechnol.* 14:1–5 (1982).
53. Wisdom, R. A., Dunnill, P., Lilly, M. D., and Macrae, A., Enzymic interesterification of fats: Factors influencing the choice of support for immobilized lipase, *Enzyme Microb. Technol.* 6:443–446 (1984).
54. Martinek, K., Levashov, A. V., Klyachko, N. L., Pantin, V. I., and Brezin, I. V., Catalysis by water-soluble enzymes entrapped into reverse micelles of surfactants in organic solvents, *Biochim. Biophys. Acta* 657:277 (1981).
55. Luisi, P. L., and Magid, L. J., Solubilization of enzymes and nucleic acids in hydrocarbon micellar solutions, *CRC Crit. Rev. Biochem.* 120(4) (1986).

7

Extractive Bioconversions in Aqueous Two-Phase Systems

Rajni Kaul and Bo Mattiasson

Chemical Center, University of Lund
Lund, Sweden

I. INTRODUCTION

The incorporation of an extractive step as a means of product separation during bioprocesses has been demonstrated to be a feasible technology for improving reactor productivity and also for reducing downstream processing costs. There is a great deal of experience in the chemical industry for performing extractions. However, it turns out that the choice of the extraction systems becomes rather limited when it comes to biotechnological applications. This is because the extractant must meet certain additional requirements when biological material is to be handled. The criteria for the selection of a suitable extractant have been listed elsewhere (1).

II. AQUEOUS TWO–PHASE SYSTEMS AS EXTRACTIVE MEDIA

Two-phase systems generated by mixing aqueous solutions of two different polymers, or a polymer and a salt in a certain range of proportions, provide a powerful method for separation of biomolecules by extraction. This extraction is based upon the surface properties of the molecule and the composition of the phase system.

Each two-phase system may be represented by a phase diagram that shows the constituent compositions at which phase separation occurs (Fig. 1) (2). All

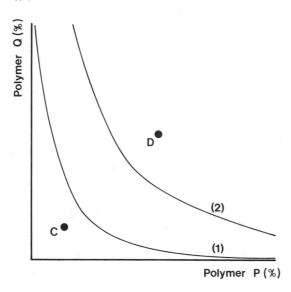

Figure 1 Mixtures of two water-soluble plymers, P and Q, represented by points above the binodial, such as point D, give two liquid phases, whereas mixtures below the binodial, such as point C, result in a homogeneous solution. The two binodials shown differ in molecular weight such that $MW2 > MW1$.

mixtures that have compositions represented by points above the line (termed the binodial) give rise to phase separation; mixtures represented by points below the line do not. The resulting phases have different compositions, with one polymer predominating in each phase. Earlier use of the aqueous two-phase systems included the separation of biomolecules on a laboratory scale, mainly for the purpose of characterization. Different subtypes of the same molecular species could have varying distribution patterns under a given set of conditions depending on their surface properties. When the difference in partition constant is too small, the separation is effected in a multistage procedure in a countercurrent distribution (CCD) machine (2), or in a chromatographic procedure with one of the polymers serving as a mobile phase and the other as a stationary phase (3). From such studies the potential of the two-phase systems as one of the means of extraction in biotechnological processes was realized, and is now supported by some large-scale work already under way on the isolation and purification of proteins. The suitability of two-phase systems for this purpose is supported by various features attached to these systems, which are outlined in this chapter.

A. Distribution Coefficient of the Product

An ideal extractant is required to have a high distribution coefficient K for the product. For aqueous two-phase systems, K is defined as the ratio of the concentration of the product in the top phase to that in the bottom phase. Large molecules and particles such as cells, phages, and high-molecular-weight DNA exhibit extreme partitioning to one of the phases (and/or the interface), proteins in general show partition coefficients in the range 0.1–10, and small ions of around 1.0. This distribution is said to be effected by several factors:

$$\ln K = \ln K_{el} + \ln K_{hphob} + \ln K_{biosp} + \ln K_{size} + \ln K_{conf}$$

where each term denotes partition contributions due to electrical charge, hydrophobicity, biospecificity, size, and conformational effects, respectively. Each of these factors, in principle, can be manipulated to enhance the selectivity of extraction of a molecule to a particular phase.

Occasionally, adsorption of compounds takes place at the interface. This must be taken into account and checked if the yield is extremely low. Therefore, several trial-and-error experiments may be required to find the suitable parameters for every extraction. The time needed for the determination of the product in the two phases is the limiting step in the development of an extraction process.

B. Biocompatibility

Water comprises 85–95% (w/w) of the polymer-polymer systems. Thus, the problems of cell toxicity that often characterize the solvent extraction procedures are not prevalent in aqueous two-phase systems.

A broad spectrum of polymers has been used for forming aqueous two-phase systems. Most of them are registered for food purposes, and some are even certified for intravenous administration to humans. It has been possible to culture microbial cells in the presence of the polymers (4). Even the labile mammalian cells have been subjected to partitioning in these systems (5). However, with such cells, consideration must be given to the membrane effects caused by poly(ethylene) glycol (PEG), which is generally used for making cell fusions. The stabilizing influence of the polymers on the enzymes has also been reported.

C. Resistance to Sanitation Conditions

The phase components can be sterilized without undergoing undue degradation. However, some precautions should be taken when dealing with carboydrate-derived polymers.

D. Product Recovery

The possibility of easy separation of the product from the extractant is another essential prerequisite for a viable system. From the aqueous two-phase systems, particulate matter may be separated by filtration or sedimentation from the favored phase and/or the interface.

Membrane filtration, salt precipitation, or adsorption to a suitable sorbent and subsequent elution may be performed quite easily for the recovery of proteins from the polymers (e.g., PEG). Another simple way to separate an enzyme protein from a PEG-rich phase is extraction into a new salt phase. The salt is removed by dialysis or membrane filtration (6). Alternatively, ion-exchange or affinity chromatography can be applied (7).

Extraction of small molecules becomes of interest during bioconversion. Their recovery from polymer solutions may be achieved either by membrane filtration (8) or by the application of selective sorbents (4).

E. Economics and Scaleup

Aqueous two-phase systems are characterized by an extremely low interfacial tension (0.1 dyn/cm). This property aids the extraction as very large surface areas are easily generated by gentle stirring, and equilibrium of partition is reached rapidly even for compounds of high molecular weight. The partition constant K is independent of the absolute concentration of the substance over a fairly wide range and can thus be used for the calculation of scaleup and performance (6). The laboratory data on enzyme isolations have been scaled up by a factor of several thousands without variation in the yields (9).

When scaling up of a particular system is required, the choice is more or less governed by the cost of the polymers, the amount needed for a particular system, the time for phase separation, and the possibility of recycling the polymer, especially if it is expensive.

PEG has constituted the top-phase and fractionated dextran, the bottom-phase polymer in the majority of the systems studied. The choice of the former is also favored by its rather low cost. However, despite its unique property of non-biodegradability under standard conditions, dextran is absolutely too expensive for scaling up. Hence there have been ongoing attempts to replace it with a cheaper polymer. Hydrolyzed crude dextran and starch have been used in some applications as more economical substitutes (10,11).

Recently, Reppal PES, a starch-based polymer, has been shown to behave as the dextrans in several respects (12); moreover, it meets the price requirements (about 40 times cheaper than fractionated dextran). The polymer has been modified with hydroxypropyl groups. Increasing the degree of substitution makes it non-biodegradable; however, for economic purposes, the substitution is limited such that a certain degree of biodegradability is retained. The evaluation of Reppal PES as a bottom-phase polymer is at an experimental stage. Another starch-based polymer, Aquaphase PPT, is also being evaluated as an alternative to dextran (13).

In a number of cases of protein purification in two-phase systems, a salt solution has constituted the bottom phase. PEG-salt systems have been found to be economically very attractive for enzyme isolation on a large scale (14). However, the high salt concentration prevents the use of these systems for bioconversions and for affinity partitioning.

Recycling of the polymers becomes extremely essential to reduce the costs of the polymers and of wastewater treatment. Extractions in two-phase systems are generally performed in a mixer-settler kind of reactor attached to a unit for the separation of the product from the phase components. The phase components are recycled to the reactor for subsequent extraction steps.

Time for phase separation is another important parameter, especially for extractive bioconversions. To achieve a fast phase separation, suitable physical properties, such as the viscosity and the density difference of the phase system, are required. The high intrinsic viscosity of certain polymers, like methylcellulose or polyvinyl alcohol, reduces the driving force for separation, thus limiting their application. It has been possible to increase the rate of phase separation by incorporation into a two-phase system of ferrofluids or iron oxide particles covered with one of the phase polymers and applying a magnetic field (15).

According to an economic analysis of the primary separation processes for cell debris separation, the price of the phase components was estimated as three-quarters of the cost of an extraction with aqueous two-phase systems (16). The total cost of an extraction, at best, was equal to the cost of tangential flow microfiltration. This analysis showed that the price of the polymers is nearly the only parameter that determines the economic feasibility of the method of extraction. In another evaluation (16) the cost of extraction was only one-half or two-thirds of the cost of other primary processes used for the recovery of intracellular enzymes.

III. AQUEOUS TWO-PHASE SYSTEMS FOR EXTRACTIVE BIOCONVERSIONS

The integration of extraction and production in a bioreactor increases the rate of product-inhibited processes. Several methods for in situ product recovery have been put forward for increasing the production of a wide range of chemicals and pharmaceuticals (17). Among these alternatives, extraction by solvents has been reported as the cheapest in a number of cases. However, it is generally observed that the better the solvent from the separation point of view, the higher is its toxicity to the cells (1,17).

The application of aqueous two-phase systems offers certain advantages when designing an extractive bioconversion. The biocompatibility of the phases is an important feature in this context. These systems can be regarded as soluble-immobilized systems for extractive bioconversion purposes: the biocatalyst is temporarily immobilized in the droplets of the phase system. This dynamic nature of immobilization results in short diffusional distances for substrates and products and, hence, high mass transfer, which is especially important in reducing microenvironmental product inhibition.

Another distinguishing feature of these systems is that it is easier to control a reaction involving a number of enzymes compared to the insolubilized systems; for example, it is possible to add more of the labile biocatalyst during continuous operation.

If the biocatalytic reaction takes place in one of the phases and the products are either evenly distributed or preferentially partitioned to the other phase, an effective extractive conversion is at hand. As mentioned earlier, microbial cells often partition exclusively to one phase, whereas the pattern of enzyme distribution is variable. Thus, to avoid the loss of enzyme partitioned to the product phase, a provision is made for retaining the biocatalyst during continuous operation (8). For low-molecular-weight products, a marked change in the distribution coefficient is difficult to achieve. Hence, the phase volumes are adjusted so that the "reaction" phase is much smaller than the other phase. The major proportion of the product is then in the phase without microbes and can be drawn off and processed appropriately. Fig. 2 shows calculated values of the percentage

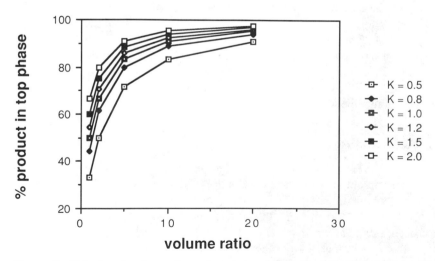

Figure 2 Calculated values of product extracted to the top phase in an aqueous two-phase system at different partition constants K and varying volume ratios.

of product extracted to the top phae in reactions with different volume ratios and different partition constants.

A. Degradation of Macromolecular or Particulate Substrates

The use of soluble enzymes has been preferred for the degradation of macro-molecular and also particulate substrates because of the diffusional barriers present in immobilized systems. As the enzymes are not recovered during such processes, the operating catalytic density is kept rather low, thereby keeping the reaction time long. The efficient mass-transfer conditions and the absence of diffusional limitations have aided the processing of high-molecular-weight sub-strates, like starch (8) and cellulose (18), in two-phase systems.

The hydrolysis of starch has been carried out in combination with an ultra-filtration step (8). A mixture of the enzymes, α-amylase and glucoamylase was used for the simultaneous liquefaction and saccharification of solid starch. The phase system consisted of 5% (w/w) 20 M PEG and 3% (w/w) crude dextran in 50 mM acetate buffer, pH 4.8. The partitioning of both the enzymes was unfavorable because they were present in the PEG-rich top phase in significant amounts. Partition constants were determined under these conditions as 0.67 and 0.51 for α-amylase and glucoamylase, respectively (19). However, introduction of the substrate to the system resulted in a α-amylase becoming adsorbed to it and subsequently improving its partition to the bottom phase. The distribution of glucoamylase, on the other hand, was not influenced substantially.

The reaction was carried out at $35°C$ in a mixer-settler reactor with a working volume of 20 L. The substrate was pumped into the reactor as a 10% (w/w) slurry and continuously converted to glucose. The phases were allowed to settle in the settling chamber, and the top phase pumped over the ultrafiltration unit. PEG and glucoamylase were retained on the upstream side of the membrane and were recirculated to the mixing chamber.

After 60 h of operation, the starch concentration was increased to 15% (w/w). The response in the yield of glucose can be seen in Fig. 3. The total yield of glucose on the basis of the amount of starch added was 94%. The process was then extended to the generation of ethanol by the addition of the yeast cells to the system after 112 h or operation time (Fig. 3). The reactor productivity was increased three times compared to conventional processes. Moreover, the stability of the enzymes was increased in the presence of the polymers.

Optimization of some of the process parameters showed that a ratio of gluco-amylase to α-amylase activities of 0.8 gave the maximal yield of glucose under the experimental conditions chosen (20). Continuous conversion of 10% (w/w) starch to glucose was then studied at $50°C$ in 0.05 M acetate buffer, pH 5.0, containing 70 ppm Ca^{2+}, the conditions providing a high reaction rate and reasonably good stability of the enzymes. Complete starch hydrolysis took place

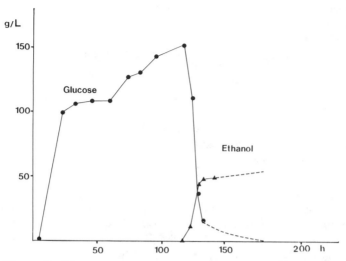

Figure 3 Glucose and ethanol produced from native starch as a function of time. The experimental conditions are given in the text.

after 18 h, and a continuous process was run for 41 h. The productivity of glucose was 13 g/L·h. An increase in bottom-phase volume was then observed as a result of the accumulation of starch. This was due to the denaturation of the enzymes (40 and 60% of the α-amylase and glucoamylase activities were lost after the first day), as the further addition of the same amount of enzyme mixture helped in continuing the reaction. During the 8 days of operation, no bacterial or fungal contamination was observed. Comparison with conventional enzymatic catalysis of starch under similar conditions showed a productivity of approximately 4.2 g/L·h. The integrated two-phase membrane separation concept gave substantial improvement. In addition, reuse of the enzymes results in a higher amount of product formed per unit of enzyme used.

The concept of combining extraction in an aqueous two-phase system with separation in ultrafiltration also protects the membrane from coming into contact with the particulate matter in the medium, thus increasing its operational life.

B. Fermentative Conversion of Small Molecules

The fermentation of monosaccharides to various solvents is normally characterized by a general inhibition phenomenon. The conversion of glucose to ethanol (19–21), acetone-butanol (22), and acetic acid (23) has been studied in aqueous two-phase systems. Phase systems with high top- to bottom-phase volume ratios have been employed. The reactor productivity has been improved in most cases compared to conventional fermentations. It was realized during repeated extractions that the buildup of by-products could be so high that they may be worth recovering by introducing a bleed procedure in the process scheme.

In the case of glucose fermentation to ethanol, repetitive batch fermentations were performed. Ethanol was stripped from the top phase by distillation. The alcohol-free phase was then returned to the fermentation flask along with more glucose. After a number of cycles, the fermentative capacity went down because of the accumulation of glycerol and other nonvolatile by-products. The system was regenerated by dialyzing the broth and adding fresh yeast cells.

The fermentation of glucose to acetone-butanol by *Clostridium acetobutylicum* was studied in a 25% (w/w) PEG and 6% (w/w) dextran T40 system. The onset of solvent production was seen to be faster in the phase system than in the buffer. However, with longer incubation times, butanol concentration decreased in the phase system but remained unchanged in the buffer (Fig. 4).

There is an indication that changes in the microenvironment of the microbial cells due to the presence of nonmetabolized polymers may lead to alterations in the metabolism of the microorganism that may partly contribute to the increased solvent production in the initial phases and, in some cases, to the formation of a new product (22). The addition of PEG and dextran to a growth medium was shown to give increased initial ethanol yields as a result of a decrease in the chemical potential of water (24). The study of the effects on the physiology of immobilized cells is an area of increasing importance (25).

The choice between batch and continuous process is thus severely influenced by the physiology of the cells used. From the data obtained it is realistic to assume a far higher production rate when operating an extractive rather than batchwise production of acetone-butanol. Furthermore, by cell recycling the fraction of substrate needed for maintenance is reduced, thus leaving a larger fraction of the substrate to be converted to the desired product.

C. One-Step Bioconversion

Some of the biological transformations of industrial importance are limited because of the poor solubility of the substrate and/or product. Bioconversion using an organic solvent as an extracting phase offers advantages with respect to downstream processing but may not be a viable method in processes utilizing living cells. It was observed that in aqueous two-phase systems, the higher solubility of such compounds could be obtained in the phase that is relatively more hydrophobic. The process studied was the transformation of hydrocortisone to prednisolone using *Arthrobacter simplex* cells (4). Both the steroids were more soluble in the top phase using poly(ethylene glycol), monomethoxypoly-(ethylene glycol) MPEG), and poly(propylene glycol (PPG) and pluronic. However, to obtain the major amounts of the product in that phase, the systems

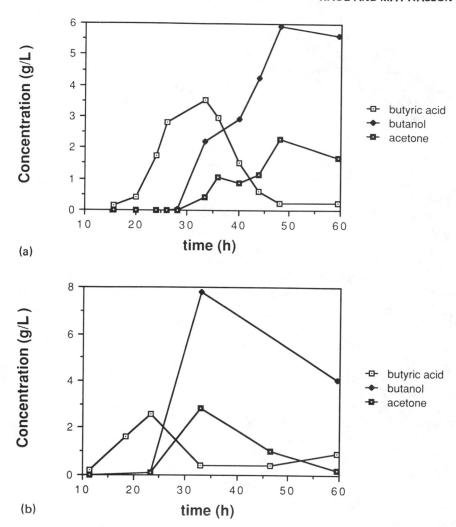

Figure 4 Product formation during fermentation of glucose by *Clostridium acetobutylicum* in batch (a) and aqueous two-phase systems (b). The concentration in the top phase is shown. Composition of the medium in g/L: glucose, 40; peptone, 10; yeast extract, 10; NH_4Cl, 0.8; Na_2HPO_4, 0.6; KH_2PO_4, 0.4; $MgSO_4 \cdot 7H_2O$, 0.2; and traces of Fe^{3+}, Ca^{2+}, Zn^{2+}, CO^{2+}, Cu^{2+}, and Mn^{2+}. The phase system consisted of medium supplied with 60 g/L Dextran T40 (Pharmacia Fine Chemicals AB, Uppsala, Sweden) and 250 g/L PEG-8000 (Union Carbide, New York), which resulted in a phase volume ratio of 6:1.

chosen had a high top- to bottom-phase volume ratio. The rate of reaction was found comparable to that in systems with organic solvent. Increasing the oxygen capacity of the medium by the use of perfluorochemical emulsion improved the reaction rate (Fig. 5). The emulsions could then be recovered as a third phase below that of dextran. The steroid product from the top phase was adsorbed on to a polymeric resin, Amberlite XAD-4, from which it was recovered into methanol. The phase polymer was recycled back to the reactor with new substrate for the repeated conversion with the cells enriched in the bottom phase. After a number of runs, it was necessary to add the nutrients to the phase system to regain the catalytic activity of the cells. It was also possible to culture the *A. simplex* in the two-phase system for long-term operations.

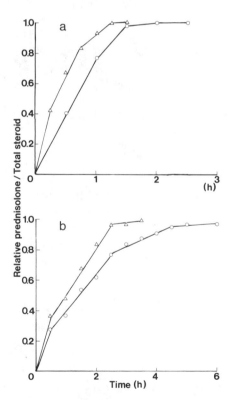

Figure 5 Transformation of hydrocortisone to prednisolone by *Arthrobacter simplex* cells in 25% PEG-8000–6% dextran T40 system in the presence (△) and (○) absence of an emulsion of a perfluorochemical. The concentration of hydrocortisone used were (a) 1 mg and (b) 3 mg/ml of the phase system. (From Ref. 4 with permission.)

The most important characteristic of this process is not the increased reaction rate but merely that the process is run almost to completion. In traditional bioconversion processes the steroids are added as crystals to the reaction mixture. Substrate dissolves, is converted, and the product crystallizes on the remaining substrate crystals. In this way mixed crystals are formed containing approximately 15% of the initial substrate shielded by the product crystallized. This phenomenon leads to expensive purification steps.

Another instance of a one-step bioconversion in a two-phase system is the deacylation of benzyl penicillin to 6-aminopenicillanic acid with penicillin acylase, which is inhibited by the acid conditions created by product formation (26). The productivity of the system was equal to that obtained with immobilized enzyme, which is an industrial process today.

D. Cell Culture

Aqueous two-phase systems can be used for culturing microorganisms to increase the productivity of the cell products, which can either be labile or toxic to the cells, by their extraction away from the vicinity of the cells. One of the earliest studies was the production of a toxin by *Clostridium tetani* in a PEG-dextran system (27). The toxin is a proteolytic enzyme capable of degrading the cell wall of the bacterium; the protoplasts formed as a result are extremely labile. By growing the bacterium in a phase system composed of 12% PEG-4000 and 2% dextran 500, with a top- to bottom-phase volume ratio of 15:1, the concentration of the protein was substantially reduced in the immediate vicinity of the cells. Thus, a more than 1000-fold increase in the total yield of toxin was obtained compared to that of the conventional medium. Apparently the phase polymers had a stabilizing effect on the protoplasts in the older cultures, which continued some metabolic activities of the cell, thus contributing to the increased synthesis.

The semicontinuous production of the enzymes, a-amylase and cellulase has been studied by culturing *Bacillus subtilis* and *Trichoderma reesei*, respectively (28,29). Some improvements in yield have been observed in both cases.

IV. PURIFICATION OF PROTEINS

The example of tetanus toxin production is one case when a protein product is extracted from the site of formation. Two-phase systems have been used extensively for the extraction of proteins from cell homogenates. In comparison, studies on extractive bioconversion are rather limited. The principle of extraction is very much the same in both cases, and one can therefore expect that this technology will be useful in extractive bioprocesses in the future.

From the K values of proteins it is evident that a maximal enrichment factor of 10 can be achieved by spontaneous distribution in the best of situations. The greatest advantage, however, is that the particulate matter and some other components that may disturb subsequent purification steps are removed quickly without centrifugation or membrane filtration steps. The technique has been successfully applied on large scale (30). Two-phase systems may also be used effectively for the concentration and simultaneous purification of labile proteins, such as those obtained from animal cell cultures. Extraction of interferon β has been carried out on a large scale with a yield of 97% (14).

Liquid ion exchangers and hydrophobic modifications have been employed in many instances for obtaining the unilateral partitioning of proteins. When affinity interactions are introduced into two-phase systems, the purification of proteins becomes much more specific and faster. The affinity extraction is set up such that the ligand and the free protein favor the opposite phases, and the ligand-protein complex partitions to the ligand phase. Generally, the ligand must be chemically modified to have it partitioned exclusively to the phase desired. In most cases, modification of the ligand with PEG (the top-phase polymer) has been carried out. A number of coupling methods could be used for this purpose (31,32).

Modification of the ligand usually lowers the binding capacity for the protein. For analytical purposes, use has been made of modified secondary separators strongly favoring the top phase and having an affinity for the ligand, which in turn interacts with the structure to be extracted. Table 1 lists examples of the affinity extraction of proteins with the aid of such secondary separators. The use of the particulate separators simplifies the removal of polymers and the subsequent recovery of the protein (34).

V. CONCLUDING REMARKS

Aqueous two-phase systems present rather useful, if not ideal extractants, to increase the rate of product-inhibited fermentations and to minimize product degradation. A good deal of know-how is available in the industry today to scale

Table 1 Affinity Extraction of Proteins Using Secondary Separators

Separator	Protein	Reference
MPEG–*Staphylococcus aureus* carrying anti-β_2-microglobulin antibodies	β_2-Microglobulin	33
MPEG-avidin	Biotin-peroxidase	33
PEG–Sepharose beads with Cibacon blue	Alcohol dehydrogenase	34

up this technology. The absence of process complexity, the need for less elaborate equipment, and ease in scaleup make this separation technology attractive for biotechnological applications.

Extractive fermentation using two aqueous phases has not really been developed sufficiently to accurately estimate its economic potential. As stated earlier, however, the cost of the polymers should be the main determinant here. To this end, a few economically acceptable polymers are being evaluated. Consideration must also be given to the polluting effect of the polymers, which would largely determine the cost of the wastewater treatment (35). From this point of view, biodegradable polymers seem to be suitable. For example, PEG is biodegradable, although by a slow process. Potential problems may also be reduced by recycling.

ACKNOWLEDGMENTS

The financial support of the National Swedish Board for Technical Development is gratefully acknowledged.

REFERENCES

1. Mattiasson, B., and Larsson, M., Extractive bioconversions with emphasis on solvent production, *Biotechnol. Genet. Eng.* 3:137–174 (1985).
2. Albertsson, P-Å., *Partition of Cell Particles and Macromolecules*, 3rd ed. Wiley-Interscience, New York (1986).
3. Müller, W., Liquid-liquid partition chromatography of biopolymers in aqueous two-phase systems, in: *Separations Using Aqueous Phase Systems. Applications in Cell Biology and Biotechnology* (D. Fisher and I. A. Sutherland, eds.). Plenum Press, New York, pp. 381–392 (1989).
4. Kaul, R., and Mattiasson, B., Extractive bioconversion in aqueous two-phase systems. Production of prednisolone from hydrocortisone using *Arthrobacter simplex* as biocatalyst, *Appl. Microbiol. Technol.* 24:259–265 (1986).
5. Walter H., Brooks, D. E., and Fischer, D., eds., *Partitioning in Aqueous Two-Phase Systems. Theory, Methods, Uses, and Applications in Biotechnology*. Academic Press, New York (1985).
6. Kula, M-R., Extraction and purification of enzymes using aqueous two-phase systems, *Appl. Biochem. Bioeng.* 2:71–95 (1979).
7. Menge, U., and Kula, M-R., Protein recovery from top phase. Poster presented at 4th Int. Conf. on Partition in Aqueous Two-Phase Systems. Lund, Sweden, August 18–23, 1985, abstract 23.
8. Larsson, M., and Mattiasson, B., Novel process technology for biotechnological solvent production, *Chem Ind* 12:428–431 (1984).
9. Kroner, K. H., Hustedt, H., Granda, S., and Kula, M-R., Technical aspects of separation using aqueous two-phase systems in enzyme isolation processes, *Biotechnol. Bioeng.* 20:1967–1988 (1978).

10. Kroner, K. H., Hustedt, H., and Kula, M-R., Evaluation of crude dextran as phase forming polymer for the extraction of enzymes in aqueous two-phase systems in large scale, *Biotechnol. Bioeng.* 24:1015-1045 (1982).

11. Hahn-Hägerdal, B., Andersson, E., Larsson, M., and Mattiasson, B., Extractive bioconversion in aqueous two-phase systems, in *Biochemical Engineering* III (K. Venkatasubramanian, A. Constantinidis, and W. R. Vieth, eds.), *Ann. N. Y. Acad. Sci.* 413:542-544 (1983).

12. Nilsson, H., Kjéllen, G., Ling, T. G. I., and Mattiasson, B., Reppal PES—a new starch derived polymer for applications in aqueous two-phase systems, in: *Proceedings of the 4th European Congress on Biotechnology* 1987, Vol. 2 (O. M. Neissel, R. R. van der Meer, and K. C. A. M. Luyben, eds.). Elsevier, Amsterdam, pp. 511-514 (1987).

13. Berner, S., Andersson, M., Tjerneld, F., and Johansson, G., Partition of biological macromolecules and particles in an aqueous two-phase system developed for biotechnical use, in: *Advances in Separations Using Aqueous Phase Systems in Cell Biology & Biotechnology*, 5th International Conference on Phase Partitioning, Oxford, August 1987, abstract 11.3.

14. Hustedt, H., Menge, U., and Kula, M-R., Protein recovery using two-phase systems, *Trends Biotechnol.* 3:139-144 (1985).

15. Wikström, P., Flygare, S., and Larsson, P-O., Magnetically enhanced aqueous two-phase separation, in: *Advances in Separations Using Aqueous Phase Systems in Cell Biology & Biotechnology*, 5th International Conference on Phase Partitioning, Oxford, August 1987, abstract 15.4.

16. Datar, R., Economics of primary separation steps in relation to fermentation and genetic engineering. *Proc. Biochem.* Feb: 19-26 (1986).

17. Roffler, S. R., Blanch, H. W., and Wilke, C. R., In situ recovery of fermentation products, *Trends Biotechnol.* 2:129-136 (1984).

18. Hahn-Hägerdal, B., Mattiasson, B., and Albertsson, P-Å., Extractive bioconversion in aqueous two-phase systems. A model study on the conversion of cellulose to ethanol, *Biotechnol. Lett.* 3:53-58 (1981).

19. Arasaratnam, V., Ph.D. Thesis, University of Jaffna, Sri Lanka (1989).

20. Larsson, M., Arasaratnam, V., and Mattiasson, B., Integration of bioconversion and downstream processing. Starch hydrolysis in aqueous two-phase system, *Biotechnol. Bioeng.* 33:758-766 (1989).

21. Kuhn, I., Alcoholic fermentation in aqeuous two-phase system, *Biotechnol. Bioeng.* 22:2393-2398 (1980).

22. Mattiasson, B., Suominen, M., Andersson, E., Häggstrom, L., Albertsson, P-Å., and Hahn-Hägerdal, B., Solvent production by *Clostridium acetobutylicum* in aqueous two-phase systems, in: *Enzyme Engineering* (I. Chibata, S. Fukui, and L. B. Wingard, Jr., eds.). Plenum, New York, vol. 6, pp. 153-155 (1982).

23. Mattiasson, B., and Hahn-Hägerdal, B., Utilization of aqueous two-phase systems for generating soluble immobilized preparations of biocatalysts, in: *Immobilized Cells and Organelles* (B. Mattiasson, ed.), CRC Press, Boca Raton, Florida, vol. 1, pp. 121-133 (1983).

24. Hahn-Hägerdal, B., Larsson, M., and Mattiasson, B., Shift in metabolism towards ethanol production in *Saccharomyces cerevisiae* using alterations of the physical-chemical microenvironment, *Biotechnol. Bioeng. Symp.* 12, 199–202 (1982).

25. de Bont, J. A. M., Visser, J., Mattiasson, B., and Tramper, J. (eds.), *Physiology of Immobilized Cells.* Elsevier, Amsterdam (1990).

26. Andersson, E., Mattiasson, B., and Hahn-Hägerdal, B., Enzymatic conversion in aqueous two-phase systems: Deacylation of benzylpenicillin to 6-aminopenicillanic acid with penicillin acylase, *Enzyme Microb. Technol.* 6: 301–306 (1984).

27. Puziss, M., and Heden, C. G., Toxin production by *Clostridium tetani* in biphasic liquid cultures, *Biotechnol. Bioeng.* 7:355–366 (1965).

28. Andersson, E., Johansson, A-C., and Hahn-Hägerdal, B., α-Amylase production in aqueous two-phase systems with *Bacillus subtilis, Enzyme Microb. Technol.* 7:333–338 (1985).

29. Persson, I., Tjerneld, F., and Hahn-Hägerdal, B., Semicontinuous cellulase production in an aqueous two-phase system with *Trichoderma reesei* Rutgers C 30, *Enzyme Microb. Technol.* 6:415–418 (1984).

30. Hustedt, H., Kroner, K. H., Stach, W., and Kula, M-R., Procedure for the simultaneous large-scale isolation pullulanase and 1,4,-β-glucan phosphorylase *Klebsiella pneumoniae* involving liquid-liquid separations, *Biotechnol. Bioeng.* 20:1989–2005 (1978).

31. Harris, J. M., Laboratory synthesis of polyethylene glycol derivatives. *J. Macromol. Sci.-Chem. Phys. Rev. Macromol. Chem Phys.* C25, 325–373.

32. Harris, J. M., Yoshinaga, K., Paley, M. S., Herati, M. R., and Upton, C. G., New activated PEG derivatives, In *Advances in Separations Using Aqueous Phase Systems in Cell Biology & Biotechnology*, 5th International Conference on Phase Partitioning, Oxford, August 1987, abstract 1.5.

33. Ling, T. G. I., and Mattiasson, B., A general study of the binding and separation in partition affinity ligand assay. Immunoassay of β_2-microglobulin, *J. Immunol. Methods* 59:327–337 (1983).

34. Mattiasson, B., and Ling, T. G. I., Efforts to integrate affinity interactions with conventional separation technologies. Affinity partition using biospecific chromatographic particles in aqueous two-phase systems, *J. Chromatogr.* 376:235–243 (1986).

35. Mattiasson, B., and Kaul, R., Use of aqueous two-phase systems for recovery and purification in biotechnology, in: *Separation, Recovery, and Purification in Biotechnology. Recent Advances and Mathematical Modeling* (J. A. Asenjo, and J. Hong, eds.). ACS Symposium Series 314, American Chemical Society, Washington, D.C. pp. 78–92 (1986).

8

Solid Sorbents Used in Extractive Bioconversion Processes

Olle Holst and Bo Mattiasson

Chemical Center, University of Lund
Lund, Sweden

I. INTRODUCTION

Among the techniques available for extractive bioconversion (Mattiasson and Larsson, 1985) or in situ product removal (Roffler et al., 1984) is the use of solid sorbents. Sorbents have been used both for removal of inhibitory products and as reservoirs for substrates of high toxicity and/or low solubility. They have also been used to adsorb readily degradable products from the reactor. Several different sorbents of various origins have been tested in biotechnological processes with promising results. The aim of this chapter is to give an overview of the technique, discussing the various sorbents, the criteria for selection of sorbents, some process concepts, and some examples of processes in which this technique has been applied, as well as some future possibilities and challenges.

II. PROCESS CONCEPTS

There are several ways to apply the technique of solid sorbents in biotechnological processes. Four different approaches are shown schematically in Fig. 1. The simplest way is to add the sorbent directly to the bioreactor (Fig. 1a). This approach can cause problems, however. Attrition of the sorbent can be severe, and adsorption of the biocatalyst can reduce the capacity of the sorbent. In addition, disturbances in the metabolism of the cells can occur if a biofilm composed of cells is formed on the sorbent surface, leading to gradients of substrate

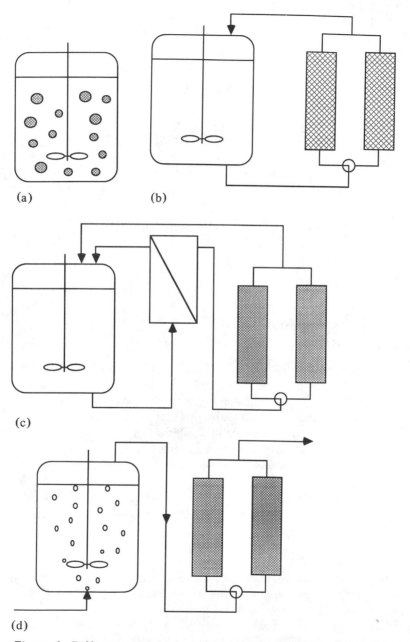

Figure 1 Different modes of applying solid sorbents in extractive bioconversion: (a) adsorption inside the fermentor; (b) adsorption in the broth but outside the fermentor; (c) adsorption in cell-free broth outside the fermentor; (d) gas stripping and subsequent adsorption outside the fermentor. For further information, see text.

and products. Furthermore, this process configuration introduces a downstream problem in that cells and sorbent particles may be difficult to separate efficiently. It is also difficult to create continuous processes based on this configuration.

A variation on this theme that appears to circumvent most of these obstacles has been presented by Wang and coworkers by introducing solid sorbents entrapped in hydrogels, such as agarose, alginate, or κ-carrageenan (Wang and Sobnosky, 1985) (Fig. 2). In this way no direct contact between the sorbent and the cells takes place, and the surface of the new particles is inert from several points of view. Drawbacks may of course appear in the form of diffusion limitations. Because of the size of the particles used, separation from cell mass was facilitated, but at the expense of intraparticle mass transfer.

The introduction of a protective layer on the sorbent reduces the risks of contact between the sorbent and the productive cells and many of the molecules in the fermentation broth. Large macromolecules do not gain access to the sorbent because of exclusion through the pores of the protective layer. The results reported on the cycloheximide process are very encouraging (Table 1) (Wang and Sobnosky, 1985).

A second approach is an off-line whole-broth treatment in which the broth is removed from the reactor and passed through a column containing the sorbent (Fig. 1b). This approach allows greater flexibility than direct addition to the reactor, as the column can easily be changed and regenerated. The problem of attrition is also avoided.

A third method is adsorption from a cell-free broth, for example one obtained by membrane filtration or centrifugal systems (Fig. 1c). This concept is preferable if cell adsorption to the surface of the sorbent is severe. Furthermore, sometimes the reaction conditions and the conditions required for an efficient adsorption are mutually exclusive. It is then advantageous to change the conditions prior to adsorption and change back before cycling the broth to the

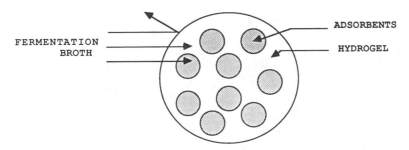

Figure 2 Entrapment of solid sorbents in hydrogels. (After Wang and Sobnosky, 1985.)

Table 1 Purity of Extracted Cycloheximide from Free and
Immobilized XAD-4 Resins

Fermentation	Solvent	Purity (%)
Shake flask (control)	Butyl acetate	39
Fermentor (control)	Butyl acetate	24
XAD-4 resin (dispersed)	Butyl acetate	54
Immobilized XAD-4 resin	Butyl acetate	87

bioreactor. This is applicable in the recovery of acids (Seevaratnam et al., in press) and to prevent the adsorption of intermediates (Nielsen et al., 1988).

As a fourth possibility gas stripping and subsequent adsorption (Fig. 1d) from the gas phase can also be employed (Walsh et al., 1983). This concept offers several advantages, as the sorbent never comes in contact with either the biocatalyst or the broth; no cell adsorption occurs, and the adsorption of nutrients and intermediates is avoided.

The combination of solid sorbents and liquid extraction using aqueous two-phase systems (see Chapter 7) as described by Kaul and Mattiasson (1986) or paraffin oil (Seevaratnam et al., in press) is also a possibility.

III. CRITERIA FOR SELECTION OF SOLID SORBENTS

Several factors must be taken into account when selecting a sorbent for the use in extractive bioconversions. These are summarized in Table 2. Some comments on the criteria are made here. In general terms, it can be stated that sorbents in direct contact with the heterogeneous broth may not be densely charged or derivatized with active, for example hydrophobic, groups. There are several examples later in this chapter in which the sorbents are based on the proper selection of the polymer backbone. Table 2 can be used as a test protocol for evaluating a sorbent for use in an extractive bioconversion. When undertaking an extractive bioconversion one of the first steps is testing for toxicity. If the sorbent is found to be compatible with the biocatalyst, its capacity for unspecific binding of essential nutrients and/or intermediates is investigated. If this is also satisfactory, adsorption isotherms are drawn up for the product of interest and related compounds. From these studies, useful information concerning selectivity, specificity, capacity, and adsorption kinetics is gained. This is then used when calculating the proper dimensions of the adsorption step in a future process.

A. Sorbent Capacity to Adsorb the Desired Compound

When applying adsorption to a solid sorbent it is obviously important that the sorbent have a high capacity for the compound to be adsorbed to minimize the volume of the sorbent used. This may be a critical point, as indicated by Wang (1983). He found it necessary to fill 25% of the bioreactor volume with adsorbent. The capacity is usually determined from the adsorption isotherm (Fig. 3).

Another very important point is that of unspecific adsorption. The lower the unspecific adsorption, the easier is the recovery of the pure product. Furthermore, a high specificity reduces the risk of eliminating essential nutrients from the medium. Adsorption of cell mass may, as stated elsewhere, cause unpredictable behavior of the system.

B. Regeneration

Desorption of the substance adsorbed should be easy. This can be done with hot gases, changes in pH or temperature, or solvent extraction. When using hot gases or liquids, the material must be heat resistant and the sorbent must be treated in such a way that no irreversible fouling due to the heat treatment occurs. Desorption with heat can be used as a means of sterilization (Groot and Luyben, 1986). In these cases it is important that the biocatalyst has not adsorbed to the sorbent or losses of catalyst will take place. It is also important to study the adsorption capacity as a function of the number of cycles the sorbent has been used.

C. Sterilization

If the sorbent is to be used in the whole-broth treatment in connection with pure cultures, the sorbent must be sterilizable. Since only heat treatment is practical on a large scale, the sorbent must be heat stable (see earlier). Sterilization is

Table 2 Criteria for Selection of Sorbents

Capacity for adsorption of the desired compound
Adsorption of biocatalyst
Costs
Possibility of regeneration
Degradation and attrition
Adsorption of nutrients
Adsorption of intermediates and precursors
Nontoxic
Sterilizability

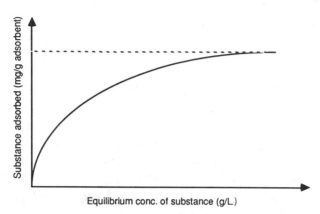

Figure 3 Adsorption isotherm.

probably not necessary if cell-free broth is treated. In this case the broth can be sterilized after the sorption step, either with heat or by filtration.

D. Toxicity

A potential advantage of solid sorbents over organic solvents as extractants is the low risk of toxicity when using polymeric material. One must, however, control two potential situations that both appear to be intoxications. The first occurs when monomers leak from the polymer. Monomers may be toxic, and true intoxication then results. This can be overcome by suitable pretreatment of the sorbent. The second case is observed as reduced or nonexistent growth. Here, however, essential nutrients have been adsorbed to the sorbent. In such cases another sorbent must be used or the sorbent can be saturated with nutrients prior to use.

E. Costs

Solid sorbents can often be reused several times, and therefore the cost is confined to the initial investment. When dealing with the production of bulk chemicals the operational costs of the solid sorbent may be the factor that determines whether the overall process is economical. For higher priced products this may not be as crucial; in these cases it is merely a matter of convenience in the process configuration and productivity. In each case, an economic evaluation must be performed.

IV. EXAMPLES OF PROCESSES IN WHICH EXTRACTIVE BIOCONVERSION IS APPLIED

A number of different solid sorbents have been applied in biotechnological processes for the removal of inhibitory compounds or as reservoirs for substrates. Both hydrophilic sorbents and ion exchangers have been used. Table 3 lists examples of sorbents used in various processes.

A. Adsorption of Toxic Products

1. Ethanol

During the last decade much attention has been paid to the fermentative production of ethanol to be used as an energy source. To reduce production costs several investigators have studied extractive bioconversion using solid sorbents to relieve product inhibition exerted by the ethanol. Among the various sorbents used for adsorption of ethanol are activated carbon (Andrews and Elkcechen, 1984; Wang et al., 1981b; Lee and Wang, 1982), silicalite (Milestone and Bibby, 1981; Lencki et al., 1983; Chung and Lee, 1985), and polymeric sorbents (Pitt et al., 1983; Lencki et al., 1983).

The capacity of the sorbent varies with the conditions, such as medium, concentration of ethanol, and temperature. In general, values around 100 mg ethanol per g sorbents are given (e.g., Malik et al., 1983; Pitt et al., 1983; Wang et al., 1981b; Milestone and Bibby, 1981). Using this approximate capacity it is possible to estimate the amount of sorbent required to reduce the ethanol content to a certain level. Assume a 100 m^3 fermentor and an initial ethanol concentration of 80 g/L. If this concentration is to be reduced to 40 g/L, then 4000 kg ethanol must be removed. Using the capacity given (0.1 g/g), 40,000 kg sorbent must be used. If the sorbent is to be added to the fermentor a significantly larger fermentor must be used, a fact that would decrease the volumetric productivity. An external loop with an adsorption step and a following desorption step is probably preferred.

As an alternative to adsorption from the liquid phase, adsorption from the gas phase has been suggested (Walsh et al., 1983). The ethanol is stripped from the fermentation broth together with carbon dioxide and water and is then adsorbed onto activated carbon. The activated carbon is desorbed using a carrier gas dried on cellulose and subsequently the ethanol is condensed from the carrier gas. An advantage of this system is that neither substrate nor cells are adsorbed onto the sorbent, as contact never occurs. At the present state of development there appears to be more realistic extractive processes for the production of ethanol (see Chapter 13).

2. Acetone and Butanol

Fermentative production of acetone-butanol-ethanol (ABE) with *Clostridium acetobutylicum* is inhibited by butanol concentrations as low as 1-2%. To improve the productivity of solvents and thereby the economy of the process, continuous product removal by solid sorbent can be applied. Various sorbents, both inorganic and organic, have been tested for application to the ABE process. Milestone and Bibby (1981) investigated the possibility of using silicalate for the adsorption of various alcohols. Butanol was concentrated from 0.5 to 98% (w/v)

Table 3 Examples of Solid Sorbents Used in Extractive Bioconversion

Sorbent	Microorganism	Substance	Reference
Activated carbon	*Saccharomyces cerevisiae*	Ethanol	Andrews and Elkcechen (1984)
Activated carbon	*S. cerevisiae*	Ethanol	Lee and Wang (1982)
Activated carbon	*S. cerevisiae*	Ethanol	Wang et al. (1981b)
Activated carbon	*Clostridium beyerincki*	Butanol-isopropanol	Groot and Luyben (1986)
Activated carbon	"Microbial film"	Organic wastewater	Andrews and Tien (1981)
Activated carbon	*S. cerevisiae*	Ethanol	Walsh et al. (1983)
Activated carbon	*Candida* sp., *Pseudomonas* sp.	Phenol	Erhardt and Rehm (1985)
Activated carbon	*Matricaria chamomilla*	Coniferyl aldehyde	Knoop and Beiderbeck (1983)
Bonopore; XAD-4 and 7	*Clostridium acetobutylicum*	Butanol	Nielsen et al. (1988), Larsson et al. (1984)
XAD-4	*Arthrobacter simplex*	Steroids	Kaul and Mattiasson (1986)
XAD-4 and 7	*S. cerevisiae*	Ethanol	Lencki et al. (1983)
XAD-2, 4, and 8	*C. beyerincki*	Butanol-isopropanol	Groot and Luyben (1986)
XAD-2, 4, and 7	*Streptomyces griseus*	Cycloheximide	Wang (1983); Wang et al. (1981a)
XAD-7	*Monascus* sp.	Red pigment	Evans and Wang (1984)
XAD-7	*Cinchona ledgeriana*	Anthraquinones	Robins and Rhodes (1986)
XAD-7	*Catharanthus roseus*	Indole alkaloids	Payne et al. (1988)
XAD-1180	*Myxococcus virescens*	Myxovirescin A	Hecht et al. (1987)
Amberlite IRA-400	*Corynebacterium renale*	Salicylic acid	Kishen Tangu and Ghose (1981)
Amberlite IRA-400	*Pseudomonas aeruginosa*	Salicylic acid	Tone et al. (1968); Kitai et al. (1968)
Silicone	*Pseudomonas testosteroni*	Steroids	Bhasin et al. (1976)
LIChroprep	Plant cells	Terpenes, etc.	Becker et al. (1984)
Unknown resin	*Pelargonium fragrans*	Monoterpenes	Charlwood et al. (1987)
Zeolite	*C. acetobutylicum*	Butanol	Larsson and Mattiasson (1984)
Silicalite	*Pachysolen tannophilus*	Ethanol	Chung and Lee (1985)
Silicalite	*C. acetobutylicum*	n-Butanol	Maddox (1982)
Silicalite	—	Alcohols	Milestone and Bibby (1981)
Silicalite	*S. cerevisiae*	Ethanol	Lencki et al. (1983)

by adsorption and subsequent thermal desorption. Zeolites have also been used (Larsson and Mattiasson, 1984). Similar capacities as for silicalite were reported by Larsson et al. (1984) for a nitrated divinylbenzene-styrene copolymer when adsorbing from an aqueous solution of pure butanol.

The capacity of various polymeric sorbents for adsorption of butanol from a water solution was between 55 and 83 mg/g sorbent (Nielsen et al., 1988). These capacities decreased significantly when the sorbent was used in cell-free spent broths because of the presence of several species likely to be adsorbed. Adsorption of nutrient was found to be a serious problem when using some sorbents.

No growth or butanol formation was found in media treated with Amberlite. However, the media could be restored by adding yeast extract.

Bonopore (a copolymer of divinylbenzene and styrene) did not affect the fermentability of the medium (Table 4).

Acetic and butyric acids are formed as intermediates in the production of acetone and butanol. It is therefore important to perform the adsorption step without enriching these acids. As the adsorption of these compounds is dependent upon the pH in the solution, a pH change was successfully used to avoid adsorption of the intermediates (Fig. 4) (Nielsen et al., 1988).

3. Butanol and Isopropanol

The possibility of using sorbents to remove solvents from cultivations of *Clostridium beyerincki* was investigated by Groot and Luyben (1986). Activated carbon and polymeric resins were studied for the adsorption of butanol, isopropanol, butyric acid, and acetic acid. Adsorption isotherms are given for butanol adsorption on five different sorbents.

4. Salicylic Acid

Microbial production of salicylic acid from naphthalene by *Pseudomonas aeruginosa* was studied by Tone et al. (1968) (see also Kitai et al., 1968). They used an ion exchanger (Amberlite IRA-400) for adsorption of the salicylic acid formed. An amount 3.75 times higher than in the control was produced. When the resin was placed in a dialysis bag to prevent contact between the cells and the resin, 5.47 times the amount of salicylate compared to the control was obtained. Similar studies have been carried out by Kishen Tangu and Goshe (1981), who studied the production of salicylic acid from naphthalene using *Corynebacterium renale*. At salicylate concentrations above 10 g/L the productivity decreases and the salicylate was degraded. As a remedy for this an ion exchanger (Amberlite IRA-400) was used to adsorb the product. The final amount of salicylic acid produced was increased by a factor fo 2.1.

B. Adsorption of Products with Low Solubilities

The low solubility of many organic chemicals in aqeuous solutions has restricted the possibility of efficiently and thus economically performing transformations

Table 4 Effect on the Fermentability of the Media When Treated with Different Adsorbents as Indicated by Cell Growth (OD 620) and Butanol Production

Culture time (h)	Untreated medium		Treated with XAD-4		Treated with XAD-4 + Yeast extract		Treated with Bonopore	
	OD 620	BuOH (g/L)	OD 620	BuOH (g/L)	OD 620	BuOH (g/L)	OD 620	BuOH (g/L)
0	0.08		0.06		0.09		0.07	
12	0.15	0.0	0.07	0.0	0.20	0.0	1.26	0.0
18	2.28	0.0	0.06	0.0	2.19	0.0	4.43	0.0
35	9.20	6.4	0.07	0.0	8.60	3.8	6.81	5.8
41	9.08	5.9	0.07	0.0	8.83	3.6	8.36	4.2

Source: After Nielsen et al., 1988.

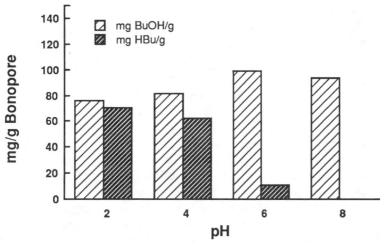

Figure 4 Adsorption of butyric acid (HBu) and butanol (BuOH) on Bonopore at different pH values. (After Nielsen et al., 1988.)

using biocatalysts that are based on viability of the cells. In other cases the addition of a volume fraction of organic solvent has been used to improve substrate solubility and thus the concentration of the substrate to be converted. Most experience has so far been gathered from steroid bioconversions. Bio-organic synthesis can be expected to grow in importance (see Laane et al., 1987).

1. Steroid Conversions

Various organic solvents, such as methanol and ethyl or butyl acetate, have been used. However, these solvents normally affect the microorganism in a negative way. As a remedy for this, Bhasin and coworkers (1976) investigated the possibility of using a solid sorbent as a reservoir for the steroids while running the conversion in an aqueous phase. They used a silicone polymer [poly(dimethylsiloxane)] in the conversion of testosterone to androst-4-ene-3,17-dione by *Pseudomonas testosteroni*. Comparisons were made with conversions carried out in a butyl acetate buffer system. As a result of the presence of the silicone polymer, the toxic butyl acetate was omitted, which led to a significant increase in final product concentration as well as productivity.

Kaul and Mattiasson (1986) carried out transformations of cortisol to prednisolone by *Arthrobacter simplex* in an aqueous two-phase system. The phase system contributed to higher solubility and better mass transfer (see Chapter 7). The recovery of prednisolone involved adsorption to Amberlite XAD-4 from the PEG-rich top phase. The prednisolone could then be back-extracted from the resin with methanol and the sorbent reused.

2. Plant Cell Products

In intact plants, lipophilic compounds are often sequestered in specific accumulation sites. These sites or structures are absent in undifferentiated cell cultures. Many of these lipophilic compounds, such as monoterpenes and essential oils, are more or less toxic, and because no accumulation sites exist in suspension cultures, they are only formed in very small amounts. As substitutes for these sites solid sorbents can be used as "artificial sinks" for the lipophilic toxic compounds.

Becker and coworkers (1984) studied the effect of the addition of RP-8 (LiChroprep, Merck, a modified silica gel with lipophilic C8 side chains) to tissue cultures of *Matricaria chamomilla, Pimpinella anisum,* and *Valeriana mallichii* producing various lipophilic compounds, such as terpenes. Two important effects were noted when studying the production of various compounds. First, the presence of the solid sorbent prevented the products from being degraded. Second, the removal of the products stimulated de novo synthesis.

Anthraquinone formation by *Cinchona ledgeriana* cultures were stimulated by the use of solid sorbents (Robins and Rhodes, 1986). Several different polymeric adsorbents were tested, and Amberlite XAD-7 showed the best characteristics. A 15-fold increase in volumetrc productivity was obtained. The presence of even small amounts of the sorbent decreased the cell growth. However, as indicated earlier, the productivity was not negatively affected. The specific productivity increased 25-fold when using XAD-7.

Charlwood and coworkers (1987) cultivated cells of *Pelargonium fragrans* in the presence of a solid sorbent as an artificial sink to remove toxic monoterpenes. Although the nature of the sorbent is not stated, they reported a significant increase in the monoterpene yield.

Activated carbon was used to adsorb coniferyl aldehyde excreted by *M. chamomilla* grown in suspension cultures. More than 95% of the substance was adsorbed to the activated carbon. The presence of activated carbon in the liquid hampered cell growth if the amount was too high: the amount should not exceed approximately 5% w/v. The reason for the inhibition remains uncertain. When 1 g activated carbon was added to a culture of 23 ml the increase in cell mass decreased to approximately 40% of that of the control, whereas the amount of coniferyl aldehyde produced was almost 60 times higher. At 2 g activated carbon in 23 ml no growth occurred. The continuous removal of coniferyl aldehyde from the culture liquor increased the productivity 60-fold compard to control (Knoop and Beiderbeck, 1983). In situ removal of alkaloids produced by *Catharanthus roseus* was studied by Payne et al. (1988). XAD-7 was used to adsorb ajmalicine and serpentine. Apart from higher productivity a change in the ratio between the two products was observed. This suggests the use of extractive bioconversion to direct metabolism toward a specific product.

3. "Red Pigment"

The fungus *Monascus* has been utilized for the production of fermented food in the Orient. It produces a red pigment, suitable as a food colorant. The pigment has low solubility in slightly acid broth and is primarily found intracellularly. Evans and Wang (1984) studied the production of this pigment, applying solid sorbents in free and immobilized cell cultures. Although no increased productivity was observed in free cell cultures, virtually all the pigment found in the broth was adsorbed.

When experiments were performed with immobilized cells (to mimic the solid-state fermentation used when fermenting rice for example), a dramatic increase in productivity was obtained by combining the immobilization with adsorption on solid sorbents. It was shown earlier that the production of secondary metabolites was stimulated upon cell immobilization (Brodelius et al., 1979). An explanation offered for this observation is that the restricted oxygen supply in the solid state favored secondary metabolite production (Mattiasson and Hahn-Hägerdal, 1982). Thus, when combining immobilization with solid sorbent for extractive removal of the product, very good improvements may be achieved.

It is obvious that this new area of bioconversion, the use of plant cell cultures, offers a very stimulating area for extractive bioconversions. One of the factors hampering the development to a substantial degree has been reputed to be the slow growth and the low metabolic activity of plant cell cultures. It is obvious that solid adsorbents offer solutions to the deposition problem to minimally differentiated and/or undifferentiated cells, thereby speeding productivity and facilitating product recovery.

C. Adsorption of Sensitive Products

1. Cycloheximide

The fermentative production of the antibiotic cycloheximide by *Streptomyces griseus* is subject to feedback regulation; the product is also degraded in the fermentor. To overcome this problem Wang and coworkers carried out extensive studies on the use of solid sorbents to remove the product from the broth (Wang et al., 1981a; Wang, 1983; Dykstra et al., 1988). Computer modeling was carried out to optimize the productivity of the process (Dykstra et al., 1988).

2. Myxovirescin A

Hecht and coworkers (1987) used a vortex chamber for the in situ recovery of the antibiotic myxovirescin A in continuous culture of *Myxococcus virescens*. The advantage of the vortex chamber is that, because of its small volume, no oxygen limitation occurs. The sorbent used was XAD-1180, designed especially for the recovery of antibiotics, such as cephalosporin. By using this technique they were able to increase the specific production rate almost fivefold compared to continuous culture without in situ product recovery. In this case the reason for the increased productivity was unclear; prevention of product degradation and diminished product inhibition were suggested.

Some of the plant-derived products mentioned earlier can also fit in here. It is generally difficult to make clear distinctions between different incentives for the in situ removal of a product in real life as done here.

D. Adsorption-Desorption as a Means of Controlling Substrate Dosage: Wastewater Treatment

An interesting application of solid sorbents in bioconversion processes is the use of the sorbent both as a support for microorganisms forming a biofilm and as a depot for toxic substrates. A theoretical approach to this has been presented, together with experimental data, by Andrews and Tien (1981) working with activated carbon treatment of organic wastewaters.

Degradation of phenol by microorganisms adsorbed on activated carbon was studied by Ehrhardt and Rehm (1985). The free cells of the microorganisms studied (*Pseudomonas* sp. and *Candida* sp.) were sensitive to high concentrations of phenol (above 1.5 g/L of phenol for *Pseudomonas*). However, when immobilized on activated carbon, shock loadings of 15 g/L were tolerated since the concentration rapidly decreased because of adsorption and degradation. This technique offers interesting possibilities for wastewater treatment as well as cultivation of microorganisms on toxic substrates (see next section).

V. FUTURE DEVELOPMENTS

The use of solid sorbents in combination with a bioconversion process is only in its infancy. So far, very little has been done to tailor sorbents for the process studied. This is an area in which much progress can be expected. One such example is in the area of zeolites, where the possibilities exist today to modify the properties, that is, the hydrophobicity and the pore sizes, so that selectivity can be introduced. By combining such specifically made sorbents with a reaction, the chances of favorable effects improve substantially.

Traditional chromatographic media have been modified to become mixed adsorbents. This means that they have two or more of the properties usually exploited separately; for example, there are mixed adsorbents based on ion exchange and hydrophobicity or on a group-specific ligand and a general property (e.g., ion exchange). An extreme are mixed gels produced by "molecular imprinting" (Wulff and Sarhan, 1972; Sellergren et al., 1985). The experience gained from even the simpler types of such supports is very limited, but provided work is invested, there will probably be a lot of interesting progress.

The technology of immobilization has been developed to a level at which many coupling methods are available and a broad selection of sorbents can be used. The immobilization of biocatalysts has been demonstrated successfully in several cases. However, there is a good chance for coimmobilization of biocatalyst and some groups or compounds that will control substrate and product concentration around the biocatalyst. This has been demonstrated for poorly soluble substrates (Kaul et al., 1986) and also in the environmental work of phenol degradation (Erhardt and Rehm, 1985).

Most studies so far have been performed on small molecules, even if the chances of introducing specificity are far superior for macromolecular substances. Thus, all the experience available from affinity purification of proteins and carbohydrates may be exploited here. Since the costs of downstream processing for processes aiming at macromolecular products is substantial and the yields often are rather low, there is space for much improvement.

The process design in this area is very conservative, and the tendency of process integration is not yet established. One obvious case when affinity extraction would be suitable is when extracting proteases from a medium in which sensitive proteins are produced.

REFERENCES

Andrews, G. F., and Elkcechens, S., Solid adsorbents in batch fermentations, *Chem. Eng. Commun.* 29:139–151 (1984).

Andrews, G. F., and Tien, C., Bacterial film growth in adsorbent surfaces, *AIChE J* 27:396–403 (1981).

Becker, H., Reichling, J., Bisson, W., and Herold, S., Two-phase cultures—a new method to yild lipophilic secondary products from plant cell cultures, in: *Proceedings of the Third European Congress on Biotechnology,* Munich Vol. 1. Verlag Chemie, Weinheim, pp. 209–213 (1984).

Bhasin D. P., Gryte, C. C., and Studebaker, J. F., A silicone polymer as a steroid reservoir for enzyme-catalysed steroid reactions, *Biotechnol. Bioeng.* 18: 1777–1792 (1976).

Brodelius, P., Deus, B., Mosbach, K., and Zenk, M. H., Immobilized plant cells for the production and transformation of natural products, *FEBS Lett.* 103: 93–97 (1979).

Charlwood, B. V., Charlwood, K. A., and Brown, J. T., The effect of product removal on the accumulation of monoterpenes in *Pelargonium* cultures, in: *Proceedings of the Fourth European Congress on Biotechnology,* Amsterdam, Vol. 2. Elsevier, Amsterdam, p. 444 (1987).

Chung, I. S., and Lee, Y. Y., Effect of in situ ethanol removal on fermentation of D-xylose by *Pachysolen tannophilus, Enzyme Microb. Technol.* 7:217–219 (1985).

Dykstra, K. H., Li, X-M., and Wang, H. Y., Computer modeling of antibiotic fermentation with on-line product removal, *Biotechnol. Bioeng.* 32:356–362 (1988).

Erhardt, H. M., and Rehm, H. J., Phenol degradation by microorganisms adsorbed on activated carbon, *Appl. Microbiol. Biotechnol.* 21:32–36 (1985).

Evans, P. J., and Wang, H. Y., Pigment production from immobilized *Monascus* sp. utilizing polymeric resin adsorption, *Appl. Environ. Microbiol.* 47:1323–1326 (1984).

Groot, W. J., and Luyben, K. Ch. A. M., In situ product recovery by adsorption in the butanol/isopropanol batch fermentation, *Appl. Microbiol. Biotechnol.* 25:29–31 (1986).

Hecht, V., Vorlop, J., Kalbitz, H., Gerth, K., and Lehmann, J., Vortex chamber for in situ recovery of the antibiotic myxovirescin A in continuous culture, *Biotechnol. Bioeng.* 24:222–227 (1987).

Kaul, R., and Mattiasson, B., Extractive bioconversion in aqueous two-phase systems: Production of prednisolone from hydrocortisone using *Arthrobacter simplex* as catalyst, *Appl. Microbiol. Biotechnol.* 24:259–265 (1986).

Kaul, R., Adlercreutz, P., and Mattiasson, B., Coimmobilization of substrate and biocatalyst: A method for bioconversion of poorly soluble substrates in water milieu, *Biotechnol, Bioeng.* 28:1432–1437 (1986).

Kishen Tangu, S., and Ghose, T. K., Environmental manipulations in salicylic acid fermentation, *Process Biochem.* August/September: 24–27 (1981).

Kitai, A., Tone, H., Ishikura, T., and Ozaki, A., Microbial production of salicylic acid from naphthalene. II. Product inhibitory kinetics and effects of product removal on the fermentation, *J. Ferment. Technol.* 46:442–451 (1968).

Knoop, B., and Beiderbeck, R., Adsorbent culture—method for the enhanced production of secondary substances in plant suspension cultures, *Z. Naturforsch.* 38c:484–486 (1983).

Laane, C., Tramper, J., and Lilly, M. D., eds., *Biocatalysis in Organic Media.* Elsevier, Amsterdam (1987).

Larsson, M., and Mattiasson, B., Novel process technology for biotechnological solvent production, *Chem. Ind.* 428–431 (1984).

Larsson, M., Holst, O., and Mattiasson, B., Butanol fermentation using a selective adsorbent for product recovery, in: *Third European Congress on Biotechnology*, vol. 2, Munich, September 10–14, 1984 (1984).

Lee, S. S., and Wang, H. Y., Repeated fed-batch rapid fermentation using yeast cells and activated carbon extraction system, *Biotechnol. Bioeng. Symp.* 12: 221–231 (1982).

Lencki, R. W., Robinson, C. W., and Moo-Young, M., On-line extraction of ethanol from fermentation broths using hydrophobic adsorbents, *Biotechnol. Bioeng. Symp.* 13:617–628 (1983).

Maddox, I. S., Use of silicalite for the adsorption of *n*-butanol from fermentation liquors, *Biotechnol. Lett.* 4:759–760 (1982).

Malik, R. K., Ghosh, P., and Ghose, T. K., Ethanol separation by adsorption-desorption, *Biotechnol. Bioeng.* 25:2277–2282 (1983).

Mattiasson, B., and Hahn-Hägerdal, Microenvironmental effects on metabolic behaviour of immobilized cells. A hypothesis, *Eur. J. Appl. Microbiol. Biotechnol.* 16:52–55 (1982).

Mattiasson, B., and Larsson, M., Extractive bioconversions with emphasis on solvent production, *Biotechnol. Genetic Eng. Rev.* 3:137–174 (1985).

Milestone, N. B., and Bibby, D. M., Concentration of alcohols by adsorption on silicalite, *J. Chem. Technol. Biotechnol.* 31:732–736 (1981).

Nielsen, L., Larsson, M., Holst, O., and Mattiasson, B., Adsorbents for extractive bioconversion applied to the acetone-butanol fermentation, *Appl. Microbiol. Biotechnol.* 28:335–339 (1988).

Payne, G. F., Payne, N. N., Schuler, M. L., and Asada, M., In situ adsorption for enhanced alkaloid production by *Catharanthus roseus, Biotechnol. Lett.* 10: 187–192 (1988).

Pitt, W. W., Haag, G. L., and Lee, D. D., Recovery of ethanol from fermentation broths using selective sorption-desorption, *Biotechnol. Bioeng.* 25:123–131 (1983).

Robins, R. J., and Rhodes, M. J. C., The stimulation of anthraquinone production by *Cinchona ledgeriana* cultures with polymeric adsorbents, *Appl. Microbiol. Biotechnol.* 24:35–41 (1986).

Roffler, S. R., Blanch, H. W., and Wilke, C. R., In situ recovery of fermentation products, *Trends Biotechnol.* 2:129–136 (1984).

Sellergren, B., Ekberg, B., and Mosbach, K., Molecular imprinting of amino acid derivatives in macroporous polymers. Demonstration of substrate- and enantio-selectivity by chromatographic resolution of racemic mixtures of amino acid derivaties, *J. Chromatogr.* 347:1–10 (1985).

Seevaratnam, S., Holst, O., Hjorleifsdottir, S., and Mattiasson, B., Extractive bioconversion for lactic acid production using solid sorbent and organic solvent, *Bioprocess. Eng.* (in press).

Tone, H., Kitai, A., and Ozaki, A., A new method for removal of inhibitory fermentation products, *Biotechnol. Bioeng.* 10:689–692 (1968).

Walsh, P. K., Liu, C. P., Findley, M. E., Liapsis, A. I., and Siehr, D. J., Ethanol separation from water in a two-stage adsorption process, *Biotechnol. Bioeng. Symp.* (13):629–647 (1983).

Wang, H. Y., Integrating biochemical separation and purification steps in fermentation processes, *Ann. N.Y. Acad. Sci.* 413:313–321 (1983).

Wang, H. Y., and Sobnosky, K., Design of a new affinity adsorbent for biochemical product recovery, in: ACS Symp. Series 271, *Purification of Fermentation Products: Applications to Large Scale Processes* (D. LeRoith, J. Shiloach, and T. J. Leahy, eds.). pp. 123–131 (1985).

Wang, H. Y., Kominek, L. A., and Jost, J. L. On-line extraction fermentation processes, *Advan. Biotechnol.* 1:601–607 (1981a).

Wang, H. Y., Robinson, F. M., and Lee, S. S., Enhanced alcohol production through on-line extraction, *Biotechnol. Bioeng. Symp.* (11):555–565 (1981b).

Wulff, G., and Sarhan, A., Über die anwendung von enzymanalog gebauten polymeren zur racemattrennung, *Angewandte Chem.* 84:364 (1972).

9

Extractive Removal of Product by Biocatalysis

Joaquim M. S. Cabral

Instituto Superior Técnico, Universidade Técnica de Lisboa
Lisbon, Portugal

I. INTRODUCTION

Biological conversions are characterized by being performed in aqueous solutions, usually at low temperatures and pressures, and by leading to dilute product solutions and low productivities. These major drawbacks of the biotechnological processes, when compared with the traditional chemical processes, are due to low biocatalyst concentrations in the reactor, inhibition end products, and low substrate and/or product solubilities in the aqueous bioconversion media.

The technological approaches to reduce these problems are based on the increase in the biocatalyst density using immobilized biocatalyst preparations and the continuous removal of the product(s) formed, as well as maintenance of the optimal operational conditions for the biocatalyst.

In situ extraction of bioproducts is a technological solution used to reduce end-product inhibition by integrating the biological conversion stage with the first steps in downstream processing to optimize the entire biotechnological process.

Liquid-liquid extraction is one unit operation that effectively utilizes the partitioning of components between two immiscible solvents or the difference in the partition coefficients between components. Extractive bioconversions with nonaqueous solvents and aqueous two-phase systems have been used in the recovery of bioproducts and are reviewed in Chapters 6 and 7.

This review addresses the subject of product recovery from a biological reaction system by a novel effective extraction process of bioconversion end products utilizing liquid-liquid extraction and enzyme reaction to achieve both separation and enrichment of valuable substances. The examples discussed come mainly from solvent and organic acid production by fermentation. However, other examples are mentioned to demonstrate the potentiality of this novel separation process.

II. LIQUID-LIQUID EXTRACTIVE BIOCONVERSIONS

From the chemical viewpoint, the two most important characteristics for an extractant are a high capacity for the product and a high selectivity for the bioproduct over water. These two requirements are described by the equilibrium distribution coefficient K_d and the separation factor a, respectively. The equilibrium distribution coefficient is defined as the ratio of the weight fraction of bioproduct in the solvent phase to that in the aqueous phase, and the separation factor a, the ratio of the distribution coefficient of the bioproduct to that of the water. (See also Chapter 6.)

For the liquid-liquid extraction of fermentation products, such as ethanol, from aqueous solutions it is useful to interpret solvent capacity and selectivity through donor-acceptor concepts (1). Ethanol and water both have electron-donating and electron-accepting capabilities. Ethanol has a slightly larger donor number (20) and a lower acceptor number (37.1) than water (18 and 54.8, respectively). It can therefore be expected that Lewis acids associate preferentially with ethanol in the organic phase and provide a greater selectivity for ethanol over water than Lewis bases. Thus, for a given equilibrium distribution coefficient the order of selectivity is carboxylic acids > alcohols > esters > amines > ketones > ethers > hydrocarbons (2). Table 1 shows the properties of different organic solvents as extractants for ethanol.

The few studies (3-6) on extractive fermentation of ethanol were carried out mainly using alcohols and esters as solvents and did not describe the use of water-immiscible organic acids for the same purpose. Cabral et al. (2) studied in situ ethanol recovery with organic acids as extractants. These compounds may also allow the extraction of alcohols with chemical reaction. On the other hand, dilute organic acid solutions (acetic, lactic, propionic, butyric, and citric acids) can be obtained by biological processes.

The concentration of organic acids is also a matter of importance because these compounds are difficult to remove from water by conventional distillation methods.

The classic product recovery method is based on precipitation of the calcium salts of the acids upon addition of calcium hydroxide to the aqueous fermentation broth. The solid is filtered off and the cake treated with sulfuric acid,

Table 1 Properties of Extracting Solvents for Ethanol

Solvents	Distribution coefficient K_d	Separation factor a
Carboxylic acids	0.15 –1.10	15–70
Alcohols	0.2 –1.30	10–40
Esters and ketones	0.1 –0.8	10–40
Amines	0.04	5
Alkyl phosphates	0.8	10
Chlorinated hydrocarbons	0.02 –0.12	50–150
Alkanes	0.0005–0.010	up to 300
Aromatics	0.04 –0.10	40–120

leading to preferential precipitation of calcium sulfate. The free organic acid in solution is purified, evaporated, and crystallized with difficulty and in low yields.

Liquid extraction has been proposed as an alternative to the classic precipitation process for the recovery of fermentation product acids (7). The extraction of carboxylic acids can be performed by (1) solvation with carbon–bonded oxygen donor extractants, including hydrocarbons and substituted hydrocarbon solvents; (2) solvation with phosphorus-bonded oxygen donor extractants; and (3) proton transfer on ion-pair formation, the extractant being aliphatic amines.

The first two categories involve the solvation of the acid by donor bonds, which are distinguished from strong covalent bonds and from ionic interactions. The latter involves extraction with reaction. Solvation with carbon-bonded oxygen donor extractants, such as alcohols, ethers, and ketones, is weaker and less specific than that obtained with the significantly more basic donor properties of the phosphorus-bonded oxygen extractants.

For biological conversion processes an additional and limiting factor influencing the selection of the extractant is the toxicity of the solvent to the biocatalyst, whether it is a microorganism or an enzyme.

General rules for the optimization of biocatalytic systems in organic media were described by Brink and Tramper (8) and Laane et al. (9) combining the data from the literature and the solvent polarity. The solvent polarity was measured by the partition coefficient of the solvent in the octanol-water two-phase system in the form of its decadic logarithm ($\log P$), known as the Hansch parameter (9), and the Hildebrandt solubility parameter δ (8).

It was found that high biocatalytic rates can be expected when the polarity of the organic solvents is low ($\delta < \sim 8$) and its molecular weight is higher than

150. However δ and the molecular weight (MW) cannot be treated separately because they are related according to

$$\delta = \left(\frac{\rho(\Delta Hv - RT)}{MW} \right)^{0.5}$$

For many solvents this implies that δ decreases automatically with increasing MW. The solubility parameter δ is hence a rather poor measure of polarity, especially for apolar solvents, and a better correlation between biocatalytic rates and polarity was observed when the partition coefficient log P was taken as an indicator of solvent polarity, as only that property is relevant, not its molecular weight (9). It can be concluded that the biological activity is low in relatively polar solvents having a log $P < 2$, is quite variable in solvents having a log P between 2 and 4, and is high in apolar solvents having a log $P > 4$. This general phenomenon was explained by Laane et al. (9) by differences in the ability of organic solvents to distort the essential water layer around biocatalysts.

Of the commonly used organic solvents, many ($\sim$50%) are not very suitable and biocompatible. Only about 20% of the solvents tested are applicable for general use in biological conversions (Table 2).

In studies of extractive fermentation of ethanol most of the solvents were shown to be toxic to ethanol-producing microorganisms and had relatively low distribution coefficient for ethanol (6).

The biocompatibile solvents for fermentation products, such as ethanol and organic acids, usually display low distribution coefficients that may render the extractive bioconversion process noncompetitive.

Extractive bioconversions involve both the extractive fermentation processes and the biotransformations of poorly soluble substances in aqueous solutions. A large number of organic compounds of potential interest to chemical and biochemical process industries have low solubilities in aqueous solutions. Biological conversions of these compounds are limited by their low solubilities. Multiphase biocatalysis may circumvent these problems using the solvent phase as substrate reservoir and/or by removing the product. The potential advantages and disadvantages of liquid two-phase reactions are listed in Table 3.

A variety of biocatalytic reactions involving the use of an organic solvent are shown in Table 4.

III. ENZYME-CATALYZED EXTRACTIVE BIOCONVERSIONS

The introduction of enzymatic catalysis in extractive bioconversions leads to a novel in situ recovery process that can achieve both the separation and enrichment of valuable substances.

Table 2 Biocompatible Organic Solvents

Solvent	Hansch parameter log P
Alcohols	
Decanol	4.0
Undecanol	4.5
Dodecanol	5.0
Oleyl alcohol	7.0
Ethers	
Diphenyl ether	4.3
Carboxylic acids	
Oleic acid	7.9
Esters	
Pentylbenzoate	4.2
Ethyldecanoate	4.9
Butyl oleate	9.8
Dibutylphthalate	5.4
Dipentylphthalate	6.5
Dihexylphthalate	7.5
Dioctylphthalate	9.6
Didecylphthalate	11.7
Hydrocarbons	
Heptane	4.0
Octane	4.5
Nonane	5.1
Decane	5.6
Undecane	6.1
Dodecane	6.6
Tetradecane	8.8
Hexadecane	9.6

Source: After Ref. 9.

Table 3 Characteristics of Two-Liquid Phase Bioconversion

Potential advantages
   High substrate and product solubilities
   Reduction in substrate and product inhibition
   Facilitated recovery of products and biocatalyst
   High gas solubility in organic solvents
   Shift of reaction equilibrium
Potential disadvantages
   Biocatalyst denaturation and/or inhibition by organic solvent
   Increasing complexity of the reaction

Table 4 Bioconversions in Organic Media

Bioconversion	Substrate	Biocatalyst	Organic solvent	Reference
Oxidation	Cholesterol	*Nocardia* sp.	Carbon tetrachloride, toluene	10
	3β- or 17β-hydroxysteroids	β-Hydroxysteroid dehydrogenase	Ethyl or butyl acetate	11
	Propene	*Microbacterium* sp.	Hexadecane	12
	Phenols	Polyphenol oxidase	Chloroform	13
Reduction	Hydrocortisone	*Arthrobacter simplex*	Dioctylphthalate	14
	Ketosteroids	β-Hydroxysteroid dehydrogenase	Ethyl or butyl acetate, iso-octane, hexane, chloroform	15
	Bicycloheptenones	β-Hydroxysteroid dehydrogenase	Octanol	16
Hydrolysis	Beef tallow, vegetable and marine oils	*Rhizopus arrhizus*	*tert*-Butyl methyl ether and acetone	17
	D,L-phenylalanine methyl ester	Chymotrypsin	Toluene	18
Esterification	Oleic acid and alcohols	Lipase	Oleic	20
	Menthol and fatty acids	Lipase	Heptane, fatty acids	21
	Glycerol and fatty acids	Lipase	Fatty acids	22

Process	Substrate	Enzyme/Organism	Solvent	Ref.
Transesterification	Tributyrin and heptanol	Lipase	Hexane, decane, ethyl, ether, carbon tetrachloride	23
Peptide synthesis	Glycols	Lipase	Ethyl acetate or butyrate	13
	Amino acids	Thermolysin, chymotrypsin	Ethyl acetate, chloroform	24
	N-protected aspartic acid and phenylalanine methyl ester	Proteinase		25
Steroid side-chain cleavage	β-Sitosterol	Mycobacterium sp.	Cyclohexane, chloroform, soybean oil	26
Fermentation				
Ethanol	Glucose	Saccharomyces cerevisiae	Dodecanol	3
			Dibutylphthalate	27
			Tri-n-butylphosphate	6
			o-tert-Butyl phenol	28
		Saccharomyces bayanus	Oleic acid	2
Acetone, butanol	Glucose	Clostridium acetobutylicum	Oleyl alcohol,	29
			C20 Guerbet alcohol	29
Lactic acid	Glucose	Lactobacillus delbrueckii	Alamine and oleyl alcohol	30

The process consists of the production and enzymatic transformation of a biological compound with simultaneous extraction of the resulting product by an appropriate organic solvent, which may also be involved in the enzyme reaction. This process is shown schematically in Fig. 1, in which the in situ recovery of the biological product A is accomplished by liquid extraction, the organic solvent B itself a substrate of the enzymatic process that could react with the bioproduct, yielding a more valuable compound C.

The biological compound could react, enzymatically, with both water-soluble and insoluble compounds, synthesizing a product that is preferentially extracted into the organic phase.

In this process, the selection of the organic-phase system must take into account the criteria defined in the previous section. Such solvents should be biocompatible so that the enyzme involved in the modification of the biological product and the biocatalysts, such as whole cells, used in its production are not deactivated. The extractant must have a larger distribution coefficient for the modified compound than for the former bioproduct to ensure a high enrichment of the product in this process. The enzyme may be used in either the free or the immobilized form.

This novel process may be suitable for the recovery of fermentation products and separation of biological compounds (Table 5).

IV. ENZYMATIC EXTRACTION OF FERMENTATION PRODUCTS

Extractive fermentations are usually limited by the hydrophilicity characteristic of the fermentation products, which lead to relatively high organic-aqueous-phase ratios, due to the low partition coefficients of those products in the organic solvent.

Technological improvement of the fermentation product recovery can be achieved by extraction with reaction, that is, enzymatic reaction, if an appropriate enzyme is introduced in the process, synthesizing a lipophilic compound, which is easily extracted into the organic phase. This lipophilic product can be directly used or hydrolyzed in the back-extraction step, yielding a purified and concentrated fermentation product. This approach can be used to recover solvents and organic acids from fermentation broths.

A. Solvents

Ethanol and acetone-butanol fermentations are biological conversion processes limited by their end products, leading to relatively diluted product solutions, 120 and 40 g/L of ethanol and butanol, respectively.

These alcohols can be extracted with water-immiscible fatty acids or oils, which allow the enzymatic esterification or interestification of the fermentation

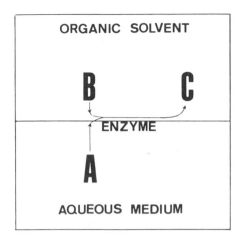

Figure 1 Enzymatic extraction of a biological product (A) with an organic solvent (B) yielding a lipophilic product (C).

products by lipases respectively (Table 5) (31). Lipases catalyze the modification of lipids, fats, and oils by their reaction with fatty acids (acidolysis, alcohols (alcholysis), or esters (transesterification), with the interchange of fatty acid groups to produce a new ester. This group of enzymes also catalyzes the hydrolysis of esters or their synthesis, depending on the water content in the reaction medium

The introduction of a lipase from the mold *Mucor miehei* in an extractive fermentation process of ethanol using water-immiscible fatty acids increases the partition coefficient of ethanol (Fig. 2) at pH 4.5 and 30°C. This enzyme exhibits a pronounced substrate specificity that increases with the chain length of the carboxylic acids. Furthermore, valeric and hexanoic acids do not undergo esterification at all.

The *M. miehei* enzyme preparation seems to display a relatively high synthetic activity, even in the presence of high water content, when hydrolysis usually prevails (20). The esterification yields of ethanol over 24 h with excess of octanoic, nonanoic, and oleic acids were 35, 43, and 65%, respecitvely.

The influence of ethanol on the rate of the ester formation at pH 4.5 and 30°C shows a typical substrate inhibition pattern (Fig. 3). The enzymatic synthesis of ethyl oleate is strongly inhibited at ethanol concentration in the aqueous phase higher than 140 g/L, and no activity was detected with 400 g/L of ethanol. Inhibition of the enzyme by increasing alkanol concentrations were also reported (20,31,32) in systems with a very low water content. However, the ethanol concentration usually found in the fermentation broth is less than 150 g/L, which allows the enzymatic esterification of ethanol without substrate

Table 5 Enzyme-Catalyzed Extractive Bioconversions

Bioconversion	Enzyme	Extractant	Enzymatic product
Solventogenic fermentation	Lipase	Fatty acids, oils	Esters
Ethanol			
Acetone, butanol			
Glycerol			Glycerides
Acidogenic fermentation	Lipase	Higher alcohols, oils	Esters
Acetic acid			
Butyric acid			
Citric acid			
Lactic acid			
Propionic acid			
Amino acid fermentation	Proteinases	Ethyl acetate	Oligopeptide
Separation of isomers	α- and β-Naphthylsulfatase	Decanol	β-Naphthol
Positional isomers			
Optical isomers			
D,L-amino acids			

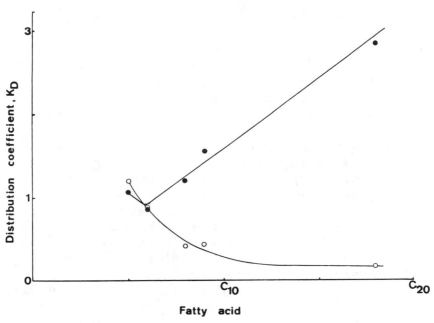

Figure 2 Distribuition coefficients for 100 g/L of initial aqueous ethanol between water-immiscible organic acids and aqueous phase (1:1 ratio): (○) physical liquid extraction and (●) liquid extraction with enzyme reaction.

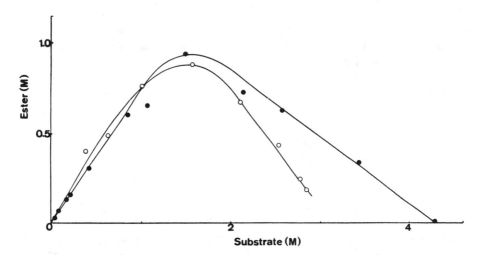

Figure 3 Effect of ethanol (●) and oleic acid (○) on the synthesis of ethyl oleate catalyzed by a *Mucor miehei* lipase.

inhibition by this lipase preparation. The esterification yields of ethanol with 1.58 M oleic acid decreased from 83 to 63% with the increase in initial ethanol concentrations in the aqueous phase from 4 to 140 g/L, respectively.

The influence of oleic acid on the enzymatic synthesis of ethyl oleate was studied at different solvent- to aqueous-phase ratios, with 110 g/L of ethanol in the aqueous medium. The effect of oleic acid is similar to that of ethanol, the rate of esterification reaction decreasing at an overall concentration of oleic acid in the reaction mixture higher than 1.58 M, which corresponds to a 1:1 solvent-to aqueous-phase ratio. The decrease in the ethyl oleate synthesis at solvent- to aqueous-phase ratios is even more pronounced than the aqueous ethanol effect (Fig. 3). The rate of reaction appears to be dependent on the oleic acid concentration, which tends either to inhibit the catalyst or to slow the rate of acid diffusion at the interface between the two phases, where the reaction seems to occur (31–33). Oleic acid was completely esterified at low concentrations (0.3–0.4 M), the yield of esterification decreasing with the increase in oleic acid phase. At a 1:1 oleic acid to aqueous-phase ratio the esterification yields of oleic acid (1.58 M) and ethanol (1.22 M) were 56 and 72%, respectively.

The time course of the esterification reaction at pH 4.5 and 30°C of ethanol (1.07 M) with oleic acid (1.58 M) in a phase ratio of 1 by *M. miehei* lipase is shown in Fig. 4. As can be seen, an equilibrium concentration of 0.71 M ethyl oleate was reached after 18 h, leading to an esterification ethanol yield of 66%. Similar conversions were obtained in the synthesis of oleic acid glycerides by *M. miehei* and *Chromobacterium viscosum* lipases in the presence of a very low (3–4%) water content (22) and ethyl oleate by *M. miehei* enzyme (20).

The ester concentration could not be increased with new enzyme addition, which confirms the equilibrium concentration obtained. An equilibrium constant of 80 was obtained for the esterification of ethanol in the biphasic system.

The effect of water content on the conversion of 40 g/L of ethanol in the aqueous phase with 1.58 M oleic acid in the biphasic system was examined by keeping the water activity of ethanol-glucose-water solution in the ranges shown in Fig. 5. The organic phase-aqueous media ratio was 1:1. Glucose was selected to reduce the water content in the aqueous phase, as this is the substrate used in the fermentation studies.

The degree of esterification achieved is inversely proportional to the water content of the system, the esterification yields of ethanol being 74 and 87% at water activities of 1 and 0.965, respectively. The higher conversion rate of ethanol at high glucose concentration is due to the best performance of ester synthesis by the *M. miehei* enzyme system with low water contents (20,22). For the experimental conditions used, the following correlation of ethyl oleate concentration in the reaction mixture with the water activity was found:

Ester (M) = $1.68 - 1.35_{aw}$

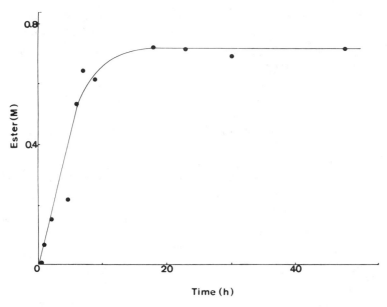

Figure 4 Effect of reaction time on the synthesis of ethyl oleate catalyzed by a *Mucor miehei* lipase from 100 g/L of initial aqueous ethanol in a 1:1 oleic acid to aqueous-phase ratio.

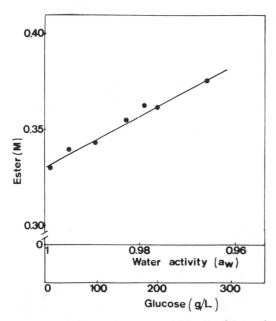

Figure 5 Effect of water activity on the synthesis of ethyl oleate calayzed by a *Mucor miehei* lipase, from 40 g/L of initial aqueous ethanol in a 1:1 oleic acid to aqueous-phase ratio.

The equilibrium constant increases with decreasing water activity, and a biphasic constant of 190 was reached at low water activity (0.965). The distribution coefficient at a high (270 g/L) glucose concentration also increases by a factor of 2.6-fold relative to the value in the absence of glucose.

The process feasibility for in situ physical and enzyme-catalyzed extractions of ethanol from a fermentation broth and the protection of cells of immobilization against solvent toxicity was tested in several extractive fermentations of solutions of 400 g/L of glucose (Fig. 6). Whole cells of *Saccharomyces bayanus* were entrapped in 4.5% w/v κ-carrageenan and oleic acid used as the extracting solvent in a 1:1 organic phase-fermentation ratio.

Immobilization of whole cells has been used to improve the ethanol fermentation performance under substrate inhibition (2) due to the higher metabolic

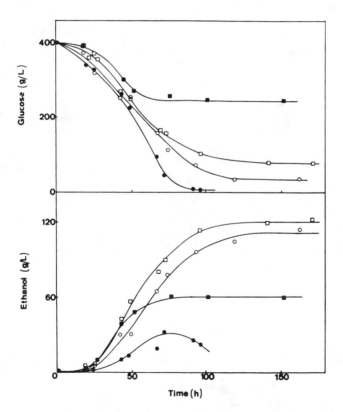

Figure 6 Fermentation of 400 g/L of glucose by *Saccharomyces bayanus*: (■) free cells; (□) κ-carrageenan–entrapped cells; (○) κ-carrageenan–entrapped cells in 1:1 oleic acid-fermentation medium ratio; and (●) κ-carrageenan–entrapped cells in 1:1 oleic acid-fermentation medium ratio in the presence of a *Mucor miehei* lipase.

efficiency than the free cell systems. Figure 6 shows this effect of entrapped cells on the fermentation of 400 g/L of glucose, the substrate residual level being 77 and 245 g/L for the κ-carrageenan–entrapped cells and their free counterparts, respecitvely (Table 6). Ethanol produced by the free cells is lower (62 g/L) than in the entrapped cells (122 g/L), which suggests that the inhibitory effect of substrate is much more pronounced than the end-product inhibition. The very different fermentation behavior may be due to the higher cell viability in the immobilized cell system than in the free cell. At high sugar concentrations, the water activity (osmolality = 3.3 Osm/kg) and oxygen supply are low, which may lead to different metabolic pathways (2,34) with accumulation of by-products other than ethanol. The gel entrapment of yeast cells creates a gradient of glucose, allowing lower microenvironmental concentrations of sugar, which would be noninhibitory to fermentation by the entrapped cells (2).

Physical extractive fermentation occurs by reducing the ethanol concentration in the aqueous phase and consequently product inhibition, which leads to the utilization of higher glucose solutions (Fig. 6 and Table 6). The residual glucose concentration in the fermentation broth was 38 g/L when a 1:1 oleic acid-fermentation medium ratio was used. As oleic acid is toxic to the cells at high concentrations (1.58 M) (2), these results suggest that cell immobilization by entrapment in κ-carrageenan protects the cell viability, probably due to steric hindrance and/or diffusional effects.

The glucose consumption increases and the level of ethanol in the aqueous phase strongly decreases (26 g/L) when a *M. miehei* lipase (86 mg/L) was added to the liquid-liquid extraction-fermentation system (Table 6).

This fermentation process was carried out with controlled pH at 4.5 to maintain the activity of the lipase, which tends to be inactive at the lower pH values usually obtained in ethanol fermentation without pH control.

Glucose was almost completely consumed without increasing the organic extractant-fermentation medium ratio, which is necessary with physical extractant (2).

The total ethyl oleate synthesized during the fermentation was 1.04 M, which corresponds to esterification yields of ethanol and oleic acid, 78.3 and 65.7%, respectively. These yields are higher than those reported previously in aqueous phases without glucose, probably due to the effect of the low water activity of the fermentation broth, which varies from 0.94 to 1 during ethanol fermentation.

In this work it was found that glucose can be metabolized by gel-entrapped yeast cells and the ethanol produced by fermentation can be esterified in situ with the extractant by a *Mucor* lipase, yielding ethyl oleate and a very low ethanol level accumulating in the fermentation broth. With this enzyme-catalyzed extractive fermentation process it is possible to synthesize esters of water-immiscible fatty acids from sugar feedstocks via the ethanol route.

Table 6 Comparison of Ethanol Fermentation Systems of 400 g/L of Glucose[a]

System	Residual glucose (g/L)	Ethanol in aqueous phase (g/L)	Glucose uptake rate (g/L·h)
A	245	61.5	1.7
B	77	122	4.0
C	38	115	4.6
D	7	22.6	5.7

[a]A, free cells; B, entrapped cells; C, entrapped cells in extractive fermentation; D, entrapped cells in lipase-catalyzed extractive fermentation.

Fats and oils may also undergo alcoholysis if they are used instead of fatty acids as extractants in the extractive fermentation of alcohols.

An extractive fermentation system may also be developed to prevent the end-product inhibition of *Clostridium acetobutylicum* by butanol, which exhibits higher toxicity to the microorganism than acetone.

The liquid-liquid extraction of butanol from the fermentation broth has been reported (29) using branched-chain alcohols of $n = 12$–20. Fatty acids can also be used in the extraction of butanol, rendering possible an esterification reaction catalyzed by lipases.

Figure 7 shows the effect of the chain length of water-immiscible fatty acids on the distribution coefficients of butanol and propanol for both physical and enzymatically catalyzed liquid extraction. The liquid-liquid extraction of butanol and propanol decreases with the chain length of the fatty acid extractant; the enzyme-catalyzed process has an optimum performance for octanoic acid, slightly decreasing with the chain length of the extractant. The enzyme-catalyzed process leads to an increase of 30-fold in the partition coefficient.

The effect of alcohol concentration on the distribution coefficients in oleic acid is shown in Fig. 8 when a *M. miehei* lipase is used to esterify ethanol, propanol, and butanol with oleic acid. The influence of butanol on the distribution coefficient shows a typical substrate inhibition pattern similar to the inhibition by ethanol.

Glycerol can be obtained by fermentation of glucose with yeast cells (35) or produced by algae (36). Glycerides can be synthesized by lipases in a two-phase system using water-immiscible fatty acids for the removal of glycerol from the fermentation or algae production broths.

Other applications of this technology are possible. For example, an important class of nonionic surfactants is composed of esters of sorbitol and long-chain fatty acids. The method just described could be used to synthesize new surfactant molecules whose composition may be selectively specified.

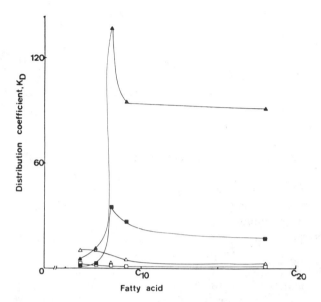

Figure 7 Distribution coefficients for 50 g/L of initial aqueous propanol (■,□) and butanol (▲,△) between water-immiscible organic acids and aqueous phase (1:1 ratio): open symbols, physical liquid extraction; closed symbols, liquid extraction with enzyme reaction.

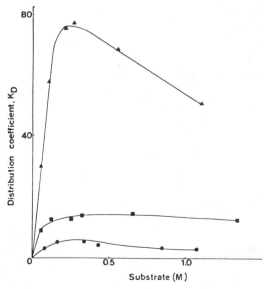

Figure 8 Effect of alcohol substrate concentration on its distribution coefficient for the enzyme-catalyzed extraction with oleic acid: (●) ethanol; (■) propanol; and (▲) butanol.

B. Organic Acids

Organic acids have been obtained as end products of both acidogenic and solventogenic fermentations. In heteofermentative processes, acetic and butyric acids are synthesized during the production of acetone and butanol by *C. acetobutylicum* and are also by far the most important inhibitors. Acetic and lactic acids are also accumulated on the ethanol fermentation of sugars by *Clostridium thermocellum*. Propionic acid is produced by different genera of bactteria, the most often used being *Propionibacterium*, with an yield of 75% from lactose. Acetic, citric, and lactic acids are commercially produced by successful fermentation processes.

Liquid extraction of these organic acids has received the most attention as a recovery process (7). The extraction of organic acids can be carried out by esterification with lipases using water-immiscible alcohols as extractants (37). Figure 9 shows the equilibria involved in this extraction-esterification process. The acid-base equilibrium of carboxylic acids in aqueous solution is represented by the dissociation constant, K_a. Only the uncharged free form of the acid, HA, is extracted into the organic phase. The distribution of the undissociated acid between the two phases is represented by the partition coefficient K_d^{HA}, and this process is referred to as physical extraction. Liquid-catalyzed esterification of the acid with an alcohol (ROH) to form an ester (RA) is associated with an equilibrium constant K_{rxn}. Under conditions of high water activity, the equilibrium concentration of the ester is expected to be low. Subsequent extraction of the ester occurs with a distribution coefficient K_d^{RA}, which is expected to be large. It was not known, however, if K_d^{RA} is sufficiently large to overcome the low equilibrium concentration of the ester and produce a significant enhancement of the overall extraction equilibrium.

The removal of propionic acid by solvent extraction and esterification in a fermentation and growth medium of *Propionibacterium freundenrichii* with oleyl alcohol was tested at pH 4.0 and 30°C using several sources of lipases. Table 7 summarizes the results obtained with several lipases after 48 h. The

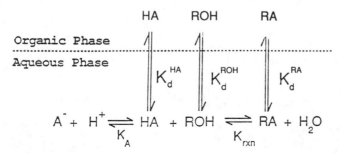

Figure 9 Underlying equilibria in coupled extraction-esterification process.

Table 7 Screening of Lipases for Oleyl Propionate Synthesis Activity

Source	Trade name (supplier)	K_d (48 h)
Control		0.80
Mucor miehei	Sp 225 (NOVO)	1.5
	Lipozyme (NOVO)	0.94
Chromobacterium viscosum	Lipase: Lp (Toyo Jozo)	2.2
Porcine pancreatic	Pancreatic Lipase (Miles)	1.0
Rhizopus sp.	Lipase N (Amano)	0.80
Penicillium sp.	Lipase G (Amano)	0.80
Aspergillus niger	Palatase A 750L (NOVO)	0.86

apparent distribution coefficients indicating each lipase's acvity for the synthesis of oleyl propionate under these conditions vary from 0.80 (the uncatalyzed value) to a maximum of 2.2 observed for the crude lipase from *Chromobacterium viscosum*. Lipases from, *Penicillium, Rhizopus* sp., and *Aspergillus niger* were not able, however, to synthesize oleyl propionate. This could be due to the higher ester synthesis activities of the lipases produced by *M. miehei* and *C. viscosum* (22).

The extraction of organic acids depends on the pH of the aqueous solution and the pK_a of the acid. The extraction of propionic acid (pK_a = 4.85) into the organic phase decreases with the increase in pH of the aqueous solution, no extraction occurring when the acid is dissociated; hence the maximum extraction occurs at low pH values (Fig. 10). With enzymatic esterification an optimum pH of 4 for extraction was observed. At low pH values the enzymatic activity is lower because of denaturation, and at high pH the acid is dissociated and cannot undergo esterification, although the enzyme displays the maximum activity for the hydrolysis (22).

Figure 11 shows the time course of both the physical extraction and esterification-coupled extraction of propionic acid with oleyl alcohol. At equilibrium, the residual propionic acid concentration was 8.3 g/L and the distribution coefficient K_dHA for the physical extraction of propionic acid was 0.80. Addition of a crude lipase from *C. viscosum* (430 mg/L of protein) to the aqueous phase strongly enhanced the extraction of propionic acid; the residual acid concentration was 1.2 g/L, and the equilibrium apparent distribution coefficient K'_d in the presence of the lipase was 12. The kinetics of the esterification process were slow, with equilibrium achieved only after approximately 50 h.

Figure 11 also presents the results of back-extraction of propionic acid from the solvent phase by enzymatic hydrolysis of oleyl propionate at pH 7.0 using a

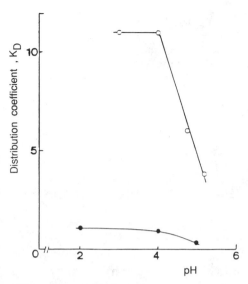

Figure 10 Effect of pH on extraction of 15 g/L of propionic acid with oleyl alcohol: (○) enzymatic extraction and (●) physical liquid extraction.

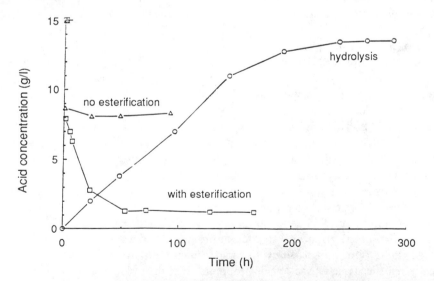

Figure 11 Time courses of propionic acid concentration during extraction into oleyl alcohol with and without enzymatic esterification and during hydrolysis. Propionic acid in aqueous phase during extraction with esterification (□), during physical extraction (△), and during hydrolysis at pH 7.0 (○).

fresh aqueous phase. Over 95% of the extracted propionic acid was recovered in the final mass balance, and the hydrolysis results confirmed the K_dHA and K'_d determinations based on high-performance liquid chromatographic (HPLC) analysis of the aqueous phase.

The enzymatic synthesis of oleyl propionate is only inhibited by high concentrations (40 g/L) of propionic acid in the aqueous phase (Fig. 12).

Water-immiscible alcohols with chain lenghts from C_4 to C_{18} were screened as extractants. The lipase from *C. viscosum* could catalyze the formation of the propionic acid ester of each of the alcohols tested. Figure 13 shows the equilibrium for straight-chain alcohols with and without enzymatic esterification for various initial concentrations of propionic acid. The equilibrium distribution coefficients (i.e., the slopes of the lines) were constant over the range of propionic acid concentrations tested for both physical and esterification-coupled extraction.

The experimental results presented in Fig. 13 serve as the basis for the numerical values in Table 8. The distribution coefficients for propionic acid decrease with the alcohol chain length, and so the least toxic solvents are the worst extractants. The results are generally in agreement with those reported in

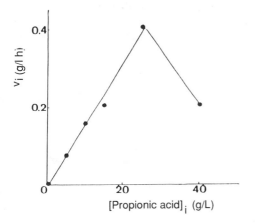

Figure 12 Effect of propionic acid on the enzymatic synthesis of oleyl propionate catalyzed by a *Cromobacterium viscosum*.

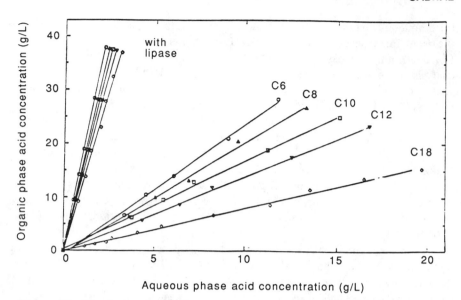

Figure 13 Physical and apparent extraction equilibria for recovery of propionic acid with water-immiscible alcohols: (O) hexanol; (△) octanol; (□) decanol; (▽) dodecanol; and (◇) oleyl alcohol.

Table 8 Extraction Properties of Water-Immiscible Alcohols by HPLC with Hydrolysis

Alcohols	$K_d\text{HA}$	K'_d	$K'_d/K_d\text{HA}$	Extracted acid esterified (%)
Butanol	3.8	18.0	4.7	80
Pentanol	3.4	17.7	5.2	81
Hexanol	2.4	17.2	7.2	86
Octanol	2.1	15.7	7.5	86
Decanol	1.7	14.5	8.5	88
Dodecanol	1.4	13.4	9.6	90
Oleyl alcohol	0.80	12.0	15.0	93

the compilation by Kertes and King (7) for the distribution coefficients of nonesterified propionic acid into n-butanol and n-pentanol.

Fortunately, the improvement in extraction produced by esterification (represented by the ratio $K'_d/K_d\text{HA}$) is greatest for the less toxic, longer chain alcohols. The use of longer chain extractants also increases the fraction extracted in the form of ester; this fraction increases with the alcohol chain length (Table 8) from 80 to 93%.

These results may be compared with those of previous work on enzymatic esterification and extraction of ethanol from a fermentation medium using water-immiscible fatty acids and a *M. meihei* lipase (31). As in the present study, the physical extraction distribution coefficient K_d was found to decrease with the acid chain length. In contrast to the weak decrease with the acid chain length observed here, however, the K'_d for ethanol extraction was found to increase with increasing acid chain length.

It was also possible to form and extract the propionic acid esters of water-miscible short-chain alcohols by blending them with a cosolvent (dodecane) to establish a water-immiscible solvent phase. Table 9 shows the propionic acid distribution coefficients for 5 and 20% solutions of ethanol in dodecane and propanol in dodecane, with and without enzymatic reaction. As in the case of the water-immiscible alcohols, ethanol and propanol could serve as substrates for the esterification of propionic acid, which was extracted principally as the ester and with an increased distribution coefficient. For the short-chain alcohols, the degree of enhancement $(K'_d/K_d\text{HA})$ was of the same order as that observed with the water-immiscible solvents (4.3–16.2 versus 4.7–15.0) and the enhancement increased with solvent chain length. The $K_d\text{HA}$ and K'_d, however, behaved differently from those obtained with the long-chain alcohols in that they both increased with chain length rather than decreased.

Table 9 Extraction and Esterification of Propionic Acid with Short-Chain Alcohol in Dodecane by HPLC with Hydrolysis

Volume % alcohol in dodecane	$K_d\text{HA}$	K'_d	$K'_d/K_d\text{HA}$	Extracted acid esterified (%)
Ethanol				
5	0.18	0.77	4.3	76
20	0.16	1.9	11.9	96
n-Propanol				
5	0.20	1.6	8.0	82
20	0.29	4.7	16.2	90

Table 10 lists a variety of organic acids that could be more effectively extracted from aqueous solutions by coupling the extraction with lipase-catalyzed esterification. For each acid, the dissociation constant pK_a, is listed along with its equilibrium distribution coefficients for extraction into oleyl alcohol with and without enzymatic esterification. The operating pH for each reaction was constrained by the need to convert the acid to the uncharged form for extraction and esterification while preserving the activity and stability of the enzyme. To avoid excessively low enzymatic reaction rates for acids with low pK_a values, the coupled enzymatic esterification and extraction of each acid was done at the pH corresponding to its pK_a (7). All other reaction conditions were as described for propionic acid.

The extraction equilibrium is substantially enhanced by esterification in each case, confirming the general applicability of the method. The degree of improvement is typically from 5- to 10-fold. The enhancement tends to be greater for the dicarboxylic acids; this may reflect multiple esterification.

The integration of an enzymatic step in the recovery of propionic acid could not be carried out in the same reaction vessel, as in the ethanol fermentation system, because of the limitations on pH for both the fermentation and esterification of propionic acid (Fig. 14). The cells are kept in the fermentation vessel by recirculation through a membrane filtration or other cell separation system. The cell-free fermentation medium is mixed with pH adjustment with the extractant in a mixer-settler device that allows the separation of the organic phase containing the enzyme-transformed organic acid and the aqueous fermentation medium, which is recycled to the fermentor.

The extraction of propionic acid can also be achieved using other water-immiscible alcohols and oils by esterification and acidolysis, respectively.

Table 10 Extraction of Carboxylic Acids with Oleyl Alcohol by HPLC and Hydrolysis

Acid	pK_a	$K_d{}^{HA}$	K'_d	$K'_d/K_d{}^{HA}$
Acetic	4.75	0.34	2.7	8.0
Citric	3.14	0.05	0.42	8.4
Formic	3.75	0.04	0.37	9.3
Fumaric	3.03	0.10	5.9	58.8
Gluconic	3.60	0.02	0.29	14.5
Itaconic	3.65	0.10	0.52	5.2
Lactic	3.86	0.02	0.17	8.5
Malic	3.40	0.02	0.20	10.0
Succinic	4.16	0.04	3.4	85.0

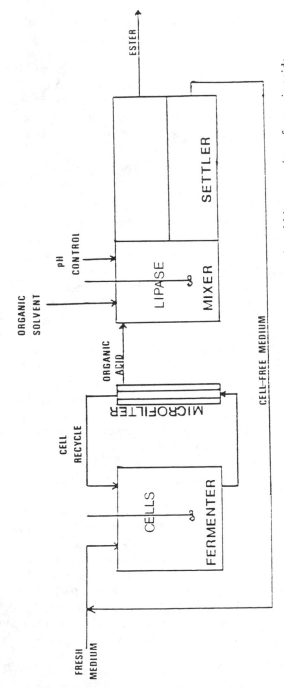

Figure 14 Continuous bioreaction system for the production and enzyme-catalyzed bioconversion of organic acids.

The model system used for the propionic acid fermentation can also be applied to other important fermentation acids, such as lactic, acetic, butyric, and citric acids (Table 5).

The use of lipases in a two-phase system to synthesize esters of both water-soluble alcohols and acids is also of importance to increase the productivity of fermentation processes. *C. thermocellum* can synthesize nearly equimolar amounts of ethanol and acetate. The esterification of these compounds by a lipase and the recovery of ethyl acetate into an organic phase could yield two beneficial results: removal of inhibitory end products from the fermentation broth and synthesis of a product easier to recover than ethanol or acetate.

C. Amino Acids

Amino acids can be obtained by both fermentation and enzymatic reactions. The major drawbacks to the biological synthesis of amino acids is the yield of the desired amino acid, especially when whole cells are used in the biological conversion. Extractive bioconversions can be used to selectively increase the molar yield of the amino acids by using an adequate multiphase system.

The liquid extraction of amino acids from aqueous solutions depends on their hydrophobicity. This extraction can be accomplished by the use of liquid emulsion membranes (38). However, with the method described previously it is possible to remove an amino acid from the bioconversion medium by reacting with other compounds, namely alcohols or other amino acids or their derivatives. A potential process is the use of proteinase for peptide synthesis from amino acids present in a fermentation broth using an organic phase as a solvent for the peptide product. The feasibility of this system depends on the type of enzyme used to promote the peptide linkage between amino acids in aqueous solutions and on the solvent for the water-insoluble peptide.

V. CONCLUDING REMARKS

This review addressed the recovery of biological products from biological conversion media, such as fermentation broth, to an organic phase using enzymatic reactions.

This novel approach enables the simultaneous production and extraction of hydrophilic compounds and their conversion into more lipophilic and valuable substances. By this technology it was shown that the fermentation productivities can improve by removal of the inhibitory end products. Enzymes, such as lipases, can synthesize ester bonds in the presence of a high water content, which renders possible their use in the integration of biological conversion and recovery processes.

However, fundamental studies on enzyme kinetics should be carried out to understand the mechanisms of action of lipases or other enzymes in the synthesis of new compounds in the presence of a high water content.

The engineering aspects of this novel technology have not yet been examined in the few examples described to illustrate the principles of this process. Mass-transfer and hydrodynamic effects and process synthesis and design should be studied to better understand the bioreaction multiphase and bioseparation processes.

Several applications of this methodology can be found in fermentation and biotransformation processes of different biological compounds, and new substances could be synthesized with selectively specified compositions for different uses in organic chemistry and the chemical and food industries.

REFERENCES

1. Munson, C. L., and King, C. J., Factors influencing solvent selection for extraction of ethanol from aqueous solutions, *Ind. Eng. Chem. Process Des. Dev.* 23:109–115 (1983).
2. Aires Barros, M. R., Cabral, J. M. S., and Novais, J. M., Production of ethanol by immobilized *Saccharomyces bayanus* in extractive fermentation system, *Biotechnol. Bioeng.* 29:1097–1104.
3. Minier, M., and Goma, G., Ethanol production by extractive fermentation, *Biotechnol Bioeng.* 24:1565–1579 (1982).
4. Mattiasson, B., and Larsson, M., Extractive bioconversions with emphasis on solvent production. *Botechnol. Genet. Rev.* 3:137–174 (1985).
5. Wang, H. Y., Robinson, F. M., and Lee, S. S., Enhanced alcohol production through on-line extraction. Biotechnol. *Bioeng. Symp.* 11:555–565 (1981).
6. Matsumura, M., and Markl, H., Application of solvent extraction to ethanol fermentation. *Appl. Microbiol. Biotechnol.* 20:371–377 (1984).
7. Kertes, A. S., and King, C. J., Extraction chemistry of fermentation product carboxylic acids. *Biotechnol. Bioeng.* 28:269–282 (1986).
8. Brink, L. E. S., and Tramper, J., Optimization of organic solvent in multiphase biocatalysis. *Biotechnol. Bioeng.* 27:1258–1269 (1985).
9. Laane, C., Boeren, S., Vos, K., and Veeger, C., Rules for the optimization of biocatalysis in organic solvents, *Biotechnol. Bioeng.* 30:81–90 (1987).
10. Ramelmeier, R. A., and Blanch, H. W., Mass transfer and cholesterol oxidase kinetics in a liquid-liquid two-phase system, *Biocatalysis* 2:97–120 (1989).
11. Cremonesi, P., Carrea, G., Spotoletti, G., and Antonini, E., Enzymatic dehydrogenation of steroids by β-hydroxysteroid dehydrogenase in a two-phase system, *Arch. Biochem. Biophys.* 159:7–10 (1973).
12. Brink, L. E. S., and Tramper, J., Facilitated mass transfer in a packed-bed immobilized-cell reactor by using an organic solvent as substrate reservoir, *J. Chem. Technol. Biotechnol.* 37:21–44 (1987).
13. Klibanov, A. M., Enzymes that work in organic solvents, *Chemtech* June: 354–359 (1986).
14. Hocknull, M. D., and Lilly, M. D., The Δ^1-dehydrogenation of hydrocortisone by *Arthrobacter simplex* in organic-aqueous two-liquid phase

environments, in: *Biocatalysis in Organic Media* (C. Laane, J. Tramper, and M. D. Lilly, eds.). Elsevier, Amsterdam, pp. 393–398 (1987).

15. Cremonesi, P., Carrea, G., Ferrara, L., and Antonini, E., Enzymatic preparation of 20-β-hydroxysteroids in two-phase systems, *Biotechnol. Bioeng.* 17: 1101–1108 (1975).

16. Leaver, J., Gartenmann, T. C. C., Roberts, S. M., and Turner, M. K., Stereospecific reductions of bicycloheptanones catalysed by 20-β-hydroxysteroid dehydrogenase, in: *Biocatalysis in Organic Media* (C. Laane, J. Tramper, and M. D. Lilly, eds.). Elsevier, Amsterdam, pp. 441–418 (1987).

17. Gancet, C., and Guignard, C., Application to ester linkage hydrolysis and synthesis in a fixed-bed reactor, in: *Biocatalysis in Organic Media* (C. Laane, J. Tramper, and M. D. Lilly, eds.). Elsevier, Amsterdam, pp. 261–266 (1987).

18. Flaschel, E., and Renken, A., Optical resolution of L-phenylalanine by enzymatic transesterification, in: *Biocatalysis in Organic Media* (C. Laane, J. Tramper, and M. D. Lilly, eds.). Elsevier, Amsterdam, pp. 375–380 (1987).

19. Kang, S. T., and Rhee, J. S., Characteristics of immobilized lipase-catalyzed hydrolysis of olive oil of high concentration in reversed phase system, *Biotechnol. Bioeng.* 33:1469–1476 (1989).

20. Gatfield, I. T., The enzymatic synthesis of esters in nonaqueous systems, *Ann N. Y. Acad. Sci.* 434:569–572 (1984).

21. Marlot, C., Langrand, G., Triantaphylides, C., and Baratti, J., Esters synthesis in organic solvent catalysed by lipases immobilized on hydrophilic supports, *Biotechnol. Lett.* 7:647–650 (1985).

22. Hoq, M. M., Tagami, H., Yamane, T., and Shimizu, S., Some characteristics of continuous glyceride synthesis by lipase in a microporous hydrophobic membrane bioreactor, *Agric. Biol. Chem.* 49:335–342 (1985).

23. Zaks, A., and Klibanov, A. M., Enzyme-catalyzed processes in organic solvents, *Proc. Natl. Acad. Sci. USA* 82:3192–3196 (1985).

24. Nakanishi, K., Kimura, Y., and Matsuno, R., Design of proteinase-catalysed synthesis of oligopeptides in an aqueous-organic biphasic system, *Bio/Technology* 4(May):452–454 (1986).

25. Ooshima, H., Mori, H., and Harano, Y., Synthesis of aspartame precursor by solid thermolysin in organic solvent, *Biotechnol. Lett.* 7:789–792 (1985).

26. Steinert, H. J., Vorlop, K. D., and Klein, J., Steroid side chain cleavage with immobilized living cells in organic solvent, in: *Biocatalysis in Organic Media* (C. Laane, J. Tramper, and M. D. Lilly, eds.). Elsevier, Amsterdam, pp. 51–64 (1987).

27. Frank, G. T., and Sirkar, K. K., Alcohol production by yeast fermentation and membrane extraction, *Biotechnol. Bioeng. Symp.* 15:621–631 (1986).

28. Honda, H., Taya, M., and Kobayashi, T., Ethanol fermentation associated with solvent extraction using immobilized growing cells of *Saccharomyces cerevisiae* and its lactose-fermentable fusant, *J. Chem. Eng. Japan* 19:268–273 (1986).

29. Ishii, S., Taya, M., and Kobayashi, T., Production of butanol by *Clostridium acetobutylicum* in extractive fermentation system, *J. Chem. Eng. Japan* 18:125–130 (1985).
30. Yabannavar, V. M., and Wang, D. I. C., New bioreactor systems using in situ solvent extraction for organic acid productin, paper presented at Biochemical Engineering V Conference, Henniker, NH, July 27 to August 1, 1986.
31. Aires Barros, M. R., Oliveira, A. C., and Cabral, J. M. S., in: *Biocatalysis in Organic Media* (C. Laane, J. Tramper, and M. D. Lilly, eds.). Elsevier, Amsterdam, pp. 185–196 (1987).
32. Knox, T., and Cliffe, K. R., Synthesis of long-chain esters in a loop reactor system using a fungal cell bound enzyme, *Process Biochem.* October:188–192 (1984).
33. Mukata, S., Kobayashi, T., and Takahasi, J., Kinetics of enzymatic hydrolysis of lipids in biphasic organic-aqueous systems, *J. Ferment. Technol.* 63:461–466 (1986).
34. Mattiasson, B., and Hahn-Hägerdal, B., Microenvironment effects on metabolic behaviour of immobilized cells, a hypothesis, *Eur. J. Appl. Microbiol. Biotechnol.* 16:52–55 (1982).
35. Kalle, G. P., Naik, S. C., and Lashkari, B. Z., Improved glycerol production from cane molasses by the sulfite process with vacuum or continuous carbon dioxide sparging during fermentation, *J. Ferment. Technol.* 63:231–237 (1985).
36. Ben-Amotz, A., and Avron, M., Glycerol and β-carotene metabolism in the halotolerant alga *Dunaliella*: A model system for biosolar energy conversion, *Trends Biochem. Sci.* November:297–299 (1981).
37. Aires Barros, M. R., Cabral, J. M. S., Willson, R-C., Hamel, J-F. P., and Cooney, C. L., Esterification-coupled extraction of organic acids: Partition enhancement and underlying reaction and distribution equilibria, *Biotechnol. Bioeng.* 34:909–915 (1989).
38. Thien, M. P., Hatton, T. A., and Wang, D. I. C., Separation and concentration of amino acids using liquid emulsion membranes, *Biotechnol. Bioeng.* 32:604–615 (1988).

10

Vacuum Fermentation

Jon Sundquist, Harvey W. Blanch, and Charles R. Wilke

University of California at Berkeley
Berkeley, California

I. INTRODUCTION

Vacuum fermentation systems remove fermentation products that are more volatile than water. Because the product to be removed must exert a greater vapor pressure than water, the application of this process is limited to a few compounds. Formic acid is the only organic acid volatile enough to be removed as a vapor. Many alcohols can be removed, although higher alcohols, such as butanols and pentanols, have pure component vapor pressures lower than water, and when dissolved at low concentrations in water, they are more volatile than water. This is due to their longer carbon chains, which increase their liquid-phase activity and thus vapor pressure.

Almost all experimental work to date with vacuum fermentations has been related to ethanol production. Ethanol is more volatile than water in mixtures up to the azeotrope, at atmospheric pressure. The highest ethanol relative volatility is at concentrations of 0–25 wt% ethanol. In addition to its high relative volatility, there has been interest in fermentative ethanol production for use as a renewable fuel.

The motivation for all extractive bioconversion processes is to alleviate inhibition caused by high product concentrations. The inhibition of ethanol and cell mass production by *Saccharomyces cerevisiae* in the presence of high ethanol concentrations has been extensively studied (1–4) and reviewed (5). Ethanol directly inhibits enzymes of the catabolic pathway (4). Evidence of inhibition

at the catabolic level is given in Fig. 1. Since the ethanol acts directly on the up-
take and conversion of glucose, ethanol productivity is always roughly proportional to the growth rate. Ethanol inhibition begins at approximately 25 g/L of
ethanol and is complete at approximately 90 g/L as shown in Fig. 2. These data
can be correlated by the following expression (4):

$$\mu = \frac{\mu_{max}S}{S + K_M} \left[1 - \frac{P}{P_{max}} \right]^n$$

where

μ_{max} = maximum specific growth rate = 0.43 h^{-1}
S = glucose concentration, g/L
K_m = Monod constant = 0.315 g/L
P = ethanol concentration, g/L
P_{max} - ethanol concentration at which growth ceases = 87.5 g/L
n = constant = 0.36

where the first term is the Monod expression for uninhibited growth and the
second term is the ethanol inhibition term.

Ethanol can also inhibit the hydrolysis of cellulose to sugars. Numerous
investigators have proposed simultaneous saccharification and fermentation
systems in which enzymatic hydrolysis and subsequent microbial conversion of
the sugars occur simultaneously in the same vessel (7–9). Ghose reported severe
inhibition of cellulases from *Trichoderma reesei* and *Aspergillus wentii* acting on
pretreated rice straw at ethanol concentrations of only 23 g/L (8). By measuring
the degree of conversion of the cellulose, it was possible to determine that
saccharification, not fermentation, was inhibited. This inhibition can be avoided
by removing ethanol as a vapor. This is accomplished by creating a vacuum in
the fermentor headspace, as shown in Fig. 3. A pressure is chosen such that the
fermentation broth boils at the fermentation temperature. The vapor produced
by boiling and subsequently removed through the vacuum compressor is more
concentrated in ethanol. This system is described in detail.

A diagram of the ethanol-water liquid-vapor equilibrium at reduced pressure
is shown in Fig 4 (10). The relative ethanol volatility is the highest at lower
ethanol concentrations, which are common in fermentation broths. As ethanol
and water vapor are removed during vacuum fermentation, nonvolatile feed components, such as salts, and nonvolatile byproducts, such as organic acids and
longer chain alcohols, accumulate in the fermentor. These species have an effect
on the relative ethanol volatility. Most of these species are ionic or polar and
depress the activity and the vapor pressure of water, resulting in an increase in
the relative ethanol volatility. This effect is much more pronounced at lower
concentrations (11). Differing quantitative results of this phenomenon have
been reported. Maiorella et al. (6) measured the ethanol volatility enhancement

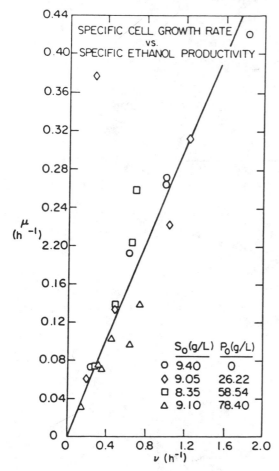

Figure 1 Relationship between the specific growth rate and specific ethanol productivity for *S. cerevisiae* subject to ethanol inhibition. S_0 is the glucose feed concentration; P_0 is ethanol feed concentration. (Data from Refs. 4 and 6.)

of several medium components in an Othmer equilibrium still and correlated the results as

$$\ln\frac{K_S}{K_0} = kC_S$$

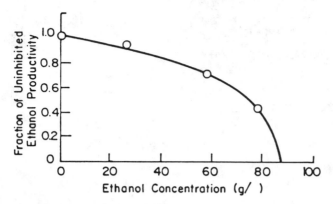

Figure 2 Inhibition of ethanol production by ethanol in *S. cerevisiae*. (Data from Ref. 4.)

where:

K_S = relative ethanol volatility in presence of added component
K_0 = relative ethanol volatility in absence of added component
C_S = concentration of added component
k = enhancement factor

The relative ethanol volatility is defined as

$$K = \frac{y_E/x_E}{y_W/x_W}$$

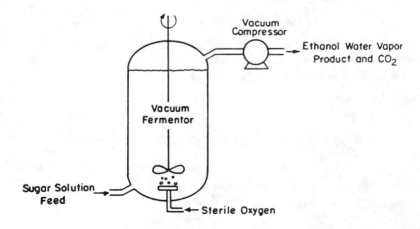

Figure 3 Vacuum fermentation.

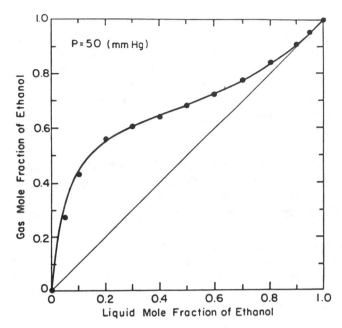

Figure 4 Ethanol-water vapor-liquid equilibrium data at 50 mmHg pressure. (From Ref. 6.)

where y_E and x_E are the vapor and liquid ethanol mole fractions, respectively, and y_W and x_W are the vapor and liquid water mole fractions, respectively.

The enhancement factors for salts ranged from 11 to 17 in 35 g/L ethanol solutions, which translates to up to a 35% enhancement of relative ethanol volatility. Yeast extract also enhanced the relative ethanol volatility to a similar degree. Fermentation broth, which was first concentrated under vacuum at 60°C and then rediluted with water and ethanol to obtain a controlled ethanol concentration, was found to have a negative enhancement factor, however (6). This result, though, may have been due to changes in the cells during concentration, allowing them to absorb ethanol. Fermentation broths treated the same way but with the cells removed demonstrated enhanced relative ethanol volatility. Roychoudury et al. (11) used a Williams equilibrium still to measure the effects of the medium components of their simultaneous saccharification and fermentation system on the relative ethanol volatility. These authors reported that the medium components—cellulases, straw, yeast cells, and nutrients—increased the relative volatility. With all components present, the relative ethanol volatility was doubled. This effect diminished as the liquid ethanol concentration increased but remained significant: 47% at 22 wt% ethanol concentration.

As great amounts of ethanol and water vapors are removed during vacuum fermentation, the concentration of the (nonvolatile) feed components, by-products, and cells increase in the fermentor, thus increasing the volatility-enhancing effect. These nonvolatile components, however, become inhibitory at high enough concentrations and thus a bleed is necessary to maintain them at a sufficiently low concentration. The salts normally found in a molasses feed have been shown to increase in inhibitory strength in the following order (12):

$$CaCl_2, (NH4)_2SO_4 > NaCl, NH_4Cl > KH_2PO_4 > MgCl_2 > MgSO_4 > KCl$$

Only 0.08 mol/L $CaCl_2$ can reduce cell mass production by 20% in continuous culture and by 80% at 0.23 mol/L. As calcium is present in molasses feedstocks at a concentration of approximately 0.058 mol/L, this ion most likely limits the extent to which the fermentation broth may become concentrated (12).

The by-products formed during ethanolic fermentation with *S. cerevisiae* are mainly organic acids and alcohols. These can be toxic through interference with the lipid cell membrane. Formic acid and acetic acid have been shown to be the most inhibitory by-products, with 80% reduction of cell mass in continuous cultures at concentrations of 2.7 and 7.5%, respectively (13). As volatile formic acid is removed overhead, acetic acid is most likely the most toxic by-product.

II. VACUUM FERMENTATION OPERATION

A. The Vacuferm

The first and simplest vacuum fermentation system was the Vacuferm, deveoped simultaneously by Ramalingham and Finn (14) and Cysewski and Wilke (15). Developmental studies were undertaken by Cysewski and Wilke, and their results (15,16) are summarized in this section. The Vacuferm system is shown in Fig. 5. This system comprises a continuous fermentor from which the vapor stream is drawn and a bleed to maintain sufficiently low concentrations of nonvolatile feed and by-product components. Cysewski and Wilke maintained a noninhibitory ethanol concentration of 3.5%, which corresponds to a pressure of 50 mmHg.

B. Fed-Batch Operation

Operation of the Vacuferm has been reported in fed-batch mode (no bleed drawn) with feeds of 10 and 33.4% glucose (15,16). Feed was added at a rate sufficient to maintain the glucose concentration between 2 and 5 g/L. The cell mass accumulation and productivity results are shown in Figs. 6 and 7. The temporal variation in ethanol productivity is due to manual intermittent adjustment of the boil-up rate used to maintain the ethanol concentration at

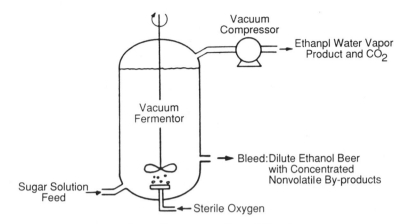

Figure 5 The Vacuferm, illustrating the cell bleed.

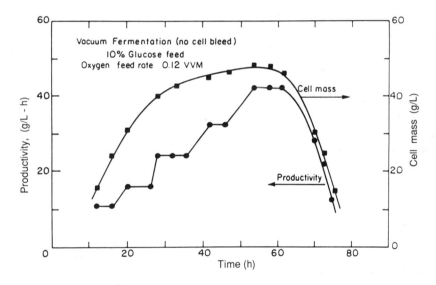

Figure 6 Cell mass productivity and concentration during fed-batch vacuum fermentation, 10% glucose feed. (From Ref. 16.)

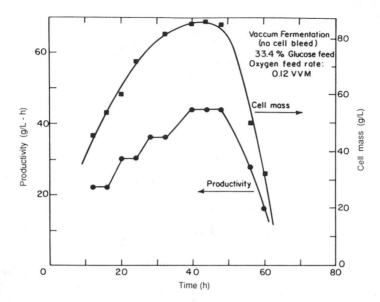

Figure 7 Cell mass productivity and concentration during fed-batch vacuum fermentation, 33.4% glucose feed. (From Ref. 16.)

approximately 3.5%. In each experiment described by these authors, an ethanol productivity of 40 g/L·h was obtained after 48 h of operation. The fermentation ended soon thereafter, however, owing to the accumulation of nonvolatile feed components and by-products.

C. Continuous Operation

Cysewski and Wilke also reported operating the Vacuferm in a continuous mode by withdrawing a liquid bleed stream from the fermentor. By measuring the cell yield at different feed-bleed ratios, it was possible to determine the extent to which the medium could be concentrated. The feed-bleed ratio is identical to the degree of concentration of feed and by-product components. These results are shown in Fig. 8. Above a concentration factor of 2.5, the cells are inhibited and cell mass decreases. By operating the continuous Vacuferm at a feed-bleed ratio of 2.4 and a dilution rate of 0.26 h^{-1}, a constant ethanol productivity of 40 g/L·h was obtained. This is identical to the peak productivity obtained in fed batch and almost six times greater than the productivity of 7 g/L·h obtained in a simple (atmospheric) continuous fermentation. The cell yield ranged from 0.055 to 0.066, approximately one-half of the cell yields of 0.10–0.12 obtained during atmospheric continuous fermentation, possibly due to the increased maintenance energy needed to grow in the presence of higher concentrations of nonvolatile components (12,13). Indeed, the specific ethanol productivity (g ethanol/g cell·h) was 0.8 h^{-1}, 38% higher than during atmospheric fermentation.

D. Continuous Operation with Cell Recycling

To further increase the volumetric productivity of vacuum fermentation, cell recycling has been employed. This involves removing the cells from the bleed stream and returning them to the fermentor. In bench-scale experiments, Cysewski and Wilke employed a cell settler to accomplish cell recycling. With cell recycling, a cell concentration of 124 g/L was obtained, an almost 12-fold increase over simple atmospheric continuous fermentation (Fig. 9). Because of the higher cell concentration, a higher dilution rate could be used and a productivity of 82 g/L·h was obtained. This system was limited only by the capacity of the settler, and potentially higher cell concentrations and productivity could otherwise be obtained. The specific ethanol productivity decreased slightly, from 0.8 to 0.66 h^{-1}. This may reflect lower yeast viability due to the extensive recycling of the cells.

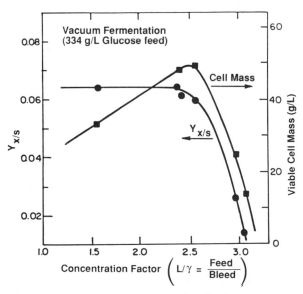

Figure 8 Cell concentration and cell yield during continuous Vacuferm operation with a 33.4% glucose feed. The reduction in cell concentration at feed-bleed ratios greater than 2.5 is due to accumulation of nonvolatile feed and by-product components.

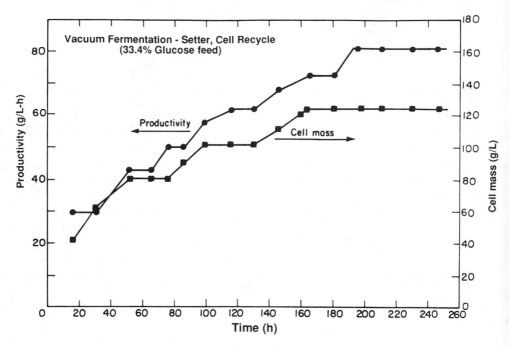

Figure 9 Cell concentration and productivity during continuous Vacuferm operation with cell recycling, with 33.4% glucose feed. Feed-bleed ratio of 2.4. The step increases in productivity and cell concentration are due to increases in feed and bleed rates. (Data from Ref. 16.)

E. Vacuferm Oxygen Requirements

S. cerevisiae generally requires oxygen for anabolic reactions, although this requirement can be met by the addition of cholesterol to the medium. In the Vacuferm, oxygen sparging at 0.12 vvm can provide the metabolic requirements. Air sparging could not, however, replace oxygen sparging under the conditions reported in Refs. 15 and 16. The low concentration of oxygen in air combined with the reduced pressure in the fermentor prevented sufficient oxygen mass transfer (16). If this system is considered for commercial application, it is likely that the nitrogen introduced during air sparging would have to be recompressed, adding to the size and cost of the vacuum compressor.

III. THE FLASHFERM

Despite the increase in productivity that can be achieved with the Vacuferm, there are a number of drawbacks that offset the economic advantages of higher volumetric productivity. These drawbacks relate to the gases (carbon dioxide and unconsumed oxygen) present in the vapor product stream (6). As already described, pure oxygen, rather than air, must be sparged to meet the oxygen requirements of the yeast, adding an additional process cost. Carbon dioxide and unconsumed oxygen leave the fermentor with the product vapor stream. Both these gases must be recompressed to the distillation column pressure by the vacuum compressor. The carbon dioxide and oxygen in the vapor stream interfere with mass transfer during vapor condensation, increasing the size of condenser needed. Finally, the gases leaving the condenser still carry approximately 10% of the ethanol product, requiring an expensive absorption step to recover this fraction.

To overcome these problems, the Flashferm has been proposed (17). In this process (Fig. 10), the fermentation is conducted at atmospheric pressure. This allows the use of air to meet the yeast oxygen requirement and allows direct venting of CO_2 gas. Ethanol-rich vapor is removed from a separate reduced-pressure flash vessel through which the broth is continually cycled. The drawbacks just mentioned can thus be circumvented. There is also less chance of contamination from leakage into the fermentor, since only the flash vessel is under vacuum (6). This process was experimentally developed and evaluated by Maiorella et al. (6). With a bench-scale apparatus, a 500 g/L glucose feed was 100% converted (18) and no adverse effects on the yeast were noticed from the constant pressure swings (6). If cell recycling is used and the fermentor cell concentration is limited to 100 g/L, a volumetric productivity of up to 123 g/L·h can be achieved, as shown by the experimental data reported in Ref. 19.

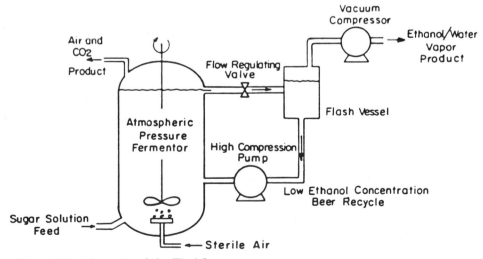

Figure 10 Schematic of the Flashferm.

IV. PROCESS IMPLICATIONS OF VACUUM FERMENTATIONS

A. Vapor Recompression

The advantage of vacuum fermentation is the reduction in ethanol inhibition, allowing higher volumetric productivity. However, higher volumetric productivity does not necessarily translate into process cost savings. An objection that has been raised against the Vacuferm is that it is too energy intensive because of the energy requirements to provide boil-up and to run the vacuum compressor (20). Although both the initial capital and operating costs of the vacuum compressor are high, this problem can be greatly reduced through the use of vapor recompression heating (21). Instead of compressing the ethanol-rich product vapors directly to the distillation column pressure, it is only necessary to compress these vapors to their dew point, at which point these heated vapors may be condensed by heat exchange with the fermentation broth, thus providing the boil-up energy. This process can be applied to both the Vacuferm and the Flashferm processes, as shown in Figs 11 and 12. To take Vacuferm as an example, optimum operating conditions, as determined by Maiorella et al. (6), call for a fermentor pressure of 78 mmHg (8.5 wt% ethanol), primary compression of the product vapors only to 360 mmHg, and then condensation of 79% of the vapors by passage through heat-exchange coils in the fermentor. A secondary compressor must be employed for the remaining vapors and gases, but the condensate can of course be cheaply pumped up to distillation column pressure. The

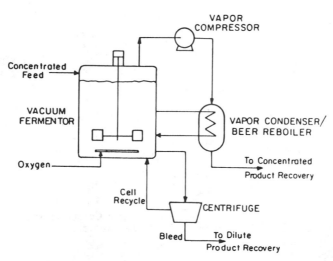

Figure 11 The Vacuferm with vapor recompression heating and cell recycling.

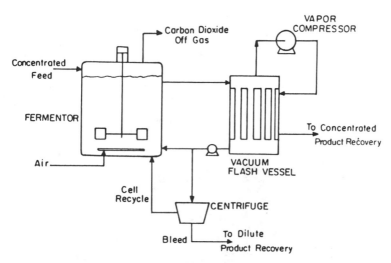

Figure 12   The Flashferm with vapor recompression heating and cell recycling.

combined capital and operating costs of the two much smaller compressors is less than for one large compressor with no intermediate vapor condensation.

B. The Frigferm

The use of vapor recompression heating reduces the vacuum compressor costs. These costs can be further reduced with the Frigferm process. This system, described by Maiorella et al. (6), uses an external heat pump to condense product vapors and provide energy for boil-up (Fig. 13). A large portion of the product vapors are condensed by cold refrigerant, which in turn is vaporized, recovering most of the heat of vaporization of the product stream. The refrigerant can then be condensed in heat-exchange coils within the flash vessel to provide heat for boil-up. The compressor required to drive the refrigeration loop is much smaller than the compressors required for flash fermentation, even if vapor recompression heating is used. This is because of the lower specific volume of the refrigerant vapors. The condensed product is cheaply pumped up to the distillation column pressure.

V. VACUUM DISTILLATION

Vacuum fermentation systems produce a more concentrated product stream compared to conventional batch or continuous fermentation. Highly concentrated product streams offer two advantages. The first is reduction if equipment size. Higher product concentration means less water diluent. Thus the initial

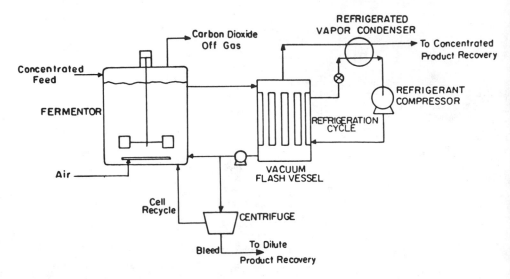

Figure 13 Schematic of the Frigferm process.

capital and operating costs of all materials-handling equipment and vessels are lower. The second advantage of highly concentrated product streams is lower separation costs, but because of the nonideal nature of the ethanol-water vapor-liquid equilibrium, distillation costs increase rapidly at levels below 6 wt% ethanol, and essentially no further savings can be obtained by increasing the product concentration above 9 wt% (19). For example, the distillation of a 5 wt% ethanol feed to 95 wt/ product requires 6.74×10^6 J/L, while concentrating 13.2 wt% ethanol to 95 wt% product requires 6.74×10^6 J/L, only 1.6% less (21). This is because energy requirements are dictated by the "pinch point" of the vapor liquid equilibrium at high school ethanol concentrations. At 96 wt% and 1 atm pressure ethanol and water form an azeotrope. The energy required for distillation is mostly determined by the amount of boil-up, which in turn is dependent on the volume of reflux. The minimum reflex ratio is determined by the slope of the equilibrium curve at the pinch point. The external reflux ratio (the amount of vapor upflow returned as reflux compared to the amount removed as purified product) is obtained from the internal reflux ratio by the expression $R_e = R_i/(1 - R_i)$, where R_e is the external reflex ratio and R_i is the internal reflex ratio (22). Figure 14 graphically illustrates the minimum reflux ratio for distillation at 1 atm pressure. The minimum external reflux ratio as calculated from the slope of the operating line is 3.316. It can be seen that the minimum feed concentration must be 2.6 mol% or 6.4 wt% increasing the feed concentration above 6.4 wt% does not affect the reflux ratio or the energy needed for separation (6).

Energy savings can be obtained if the distillation is carried out at reduced pressure. At low pressures, the ethanol-water vapor-liquid equilibrium shifts so that the reflux ratio required is not as high, as shown in Fig. 15. At a pressure of 50 mmHg, the external reflux ratio is only 1.45 and thus energy savings can be achieved. A minimum feed concentration of 7.8 mol% (17.8 wt%) is required, however. The high product concentrations produced by the vacuum fermentation process can thus result in a lowering of purification costs if vacuum distillation is employed. Vacuum distillation also complements vacuum fermentation in that the condensed vapor products need not be compressed to atmospheric pressure prior to distillation.

VI. ECONOMIC COMPARISONS

Detailed design and economic comparisons are available for most of the proposed ethanol fermentation systems (6,19). These evaluations include not only the direct fermentation costs but also the costs associated with distillation, feed

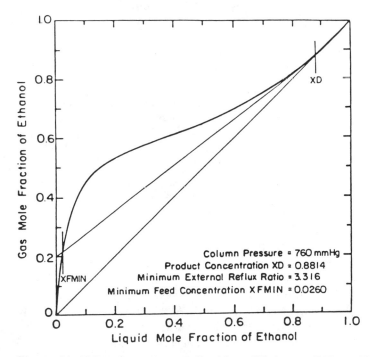

Figure 14 Ethanol-water vapor-liquid equilibrium at 760 mmHg pressure with the rectifying section operating line shown for minimum reflux conditions. (From Ref. 6.)

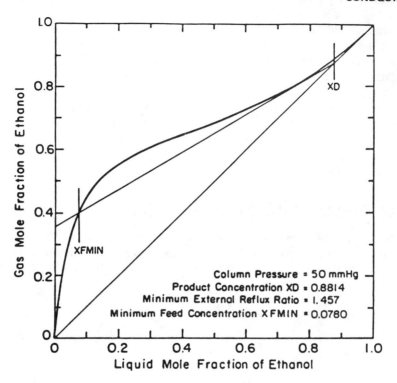

Figure 15 Ethanol-water vapor liquid equilibrium at 50 mmHg pressure, with the rectifying section operating line shown for minimum reflux conditions. (From Ref. 6.)

preparation, and stillage handling and have all been compared on an identical basis using a consistent model of the yeast metabolism. Molasses was chosen for a substrate. The results of these analyses for optimized vacuum fermentation, as well as for batch and recycling CSTR systems, are compared in Table 1 (data given in 1983 U.S. dollars). Batch fermentation is clearly the most expensive route to ethanol production. Neither Vacuferm nor Flashferm, however, offers any cost advantage over recycling CSTR. Frigferm offers a marginal 0.55 cents/L savings, which may be offset in practice by operational difficulties of this complex process. This conclusion is due, surprisingly, to a reduction in the yeast by-product credit that can be claimed. Yeast growth is reduced by high concentrations of feed and by-product concentrations in vacuum fermentation. Should the value of yeast (as cattle feed) decrease in the future, vacuum fermentation systems will become more competitive owing to lower capital and operational costs compared to recycling CSTR.

Table 1 Fermentation Process Comparison

	Purchased equipment cost				Manufacturing cost (¢/L)	Yeast credit (¢/L)	Final product cost (¢/L)
	Storage (10^5)	Fermentation (10^5)	Ethanol + stillage recovery (10^5)	Total (10^5)			
Batch	6.90	18.01	33.86	58.77	10.9	4.8	53.05
Recycling CSTR	6.90	11.31	30.21	48.42	8.64	4.8	49.06
Vacuferm	7.11	27.22	16.93	51.25	6.85	2.1	51.48
Flashferm	6.92	26.66	12.70	46.28	6.20	1.7	49.44
Frigferm	6.92	22.19	12.69	41.80	5.93	1.7	48.51

Source: From Ref. 6.

VII. OTHER APPLICATIONS OF VACUUM FERMENTATION

A. Flashferm with Thermophilic Microorganisms

Thermophilic anaerobic bacteria have received considerable attention as possible ethanol producers. *Clostridium thermohydrosulfuricum* can produce ethanol from glucose with a yield of 0.48 g/g (94% of the theoretical yield), comparable to yeasts (23). More importantly, these bacteria can ferment xylose, an abundant sugar that *S. cerevisiae* is unable to metabolize. *C. thermohydrosulfuricum* converts xylose to ethanol with a yield of 0.39 g/g (24). Other species, such as *Clostridium thermocellum* produce cellulolytic enzymes, which has prompted research involving simultaneous saccharification and fermentation (9,25). These bacteria seem to be especially suited for growth in a vacuum fermentation system. First, their high growth temperature (approximately 65°C) allows much higher vacuum pressures to be used. Pressures of 200 mmHg are sufficient to remove product vapors, compared to 50–100 mmHg required for mesophilic yeast fermentations. The lower pressures and denser vapors greatly reduce vacuum compressor costs. Second, these bacteria are much more sensitive to ethanol. Some strains can grow up to 6% ethanol, but only at a greatly reduced yield and growth rate (9). Additionally, these strains often lose their ethanol tolerance over time.

For these reasons, vacuum fermentation has been suggested as a preferred fermentation process (25). There are problems associated with this approach, however. These bacteria produce acetic and lactic acids, hydrogen, and carbon dioxide as by-products. The distribution of end products is governed by the availability of reducing power within the cells (25). If the hydrogen is allowed to escape from the fermentation broth, these cells react by depleting reducing power to produce more hydrogen, at the expense of the most reduced organic product, ethanol. With *C. thermohydrosulfuricum*, it has been shown that when the medium is allowed to be depleted of hydrogen gas by nitrogen sparging, organic acids become the major product (26). This bacterium has been grown in a flash fermentation system (Fig. 16). When a vacuum of 60 kPa (451 mmHg) was applied, the lactic acid concentration began to rise. It remained high as the pressure was gradually lowered but nevertheless remained above the vapor pressure of the broth. When the pressure was reduced to 27 kPa (203 mmHg), the vapor pressure of the 65°C broth, boiling was observed in the flash vessel and lactic acid became the major product, eventually reaching 10–12 g/L. The drop in ethanol concentration was not attributable to removal in the vapor phase. Rather, as boiling developed in the flash vessel, dissolved hydrogen was removed, causing the bacteria to redirect their reducing power to further hydrogen production at the expense of ethanol production (27). Because of the importance of dissolved hydrogen, vacuum fermentation does

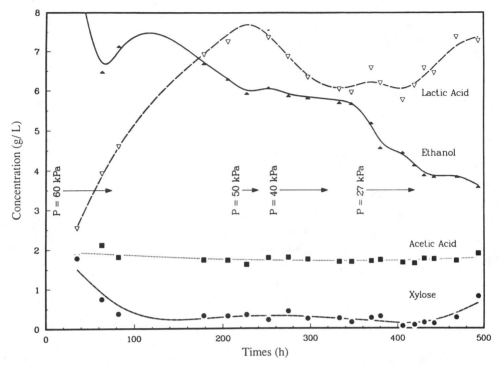

Figure 16 Flash fermentation of xylose (30 g/L feed), with *C. thermohydro-sulfuricum.* Lactic acid production rises as flash pressure is reduced to 60 kPa (451 mmHg). At 27 kPa (203 mmHg), the vapor pressure of the broth, ethanol production decreases as hydrogen is lost in the flash vessel (Data from Ref. 27.)

not appear to be easily applicable to thermophilic ethanolic fermentations by *Clostridium.*

B. Flashferm with *Zymomonas mobilis*

Flash fermentation has been successfully applied to the bacterium *Z. mobilis,* however. This bacterium, which produces ethanol at a higher yeild than yeasts, is also fairly sensitive to ethanol, although it can thrive in up to 60 g/L (28). *Z. mobilis* was grown in a flash fermentation system with step cycling of broth to the flash vessel. This operation reduces the amount of compressor running time, although it increases the size of the compressor and fermentation vessels. Maintaining the ethanol concentration at 60 g/L, the brink of severe inhibition, feeds of up to 350 g/L can easily be fermented without cell recycling. With cell recycling and a 200 g/L feed, a productivity of 80 g/L·h was obtained, with a condensate ethanol concentration of 200 g/L (29). This performance compared well with yeast ethanolic fermentation.

C. Flashferm with SSF

The step cycling flash fermentation system has also been successfully applied to simultaneous saccharification and fermentation (SSF). In a system using cellulases from *T. reesei* and *A. wentii* and the yeast *Candida acidothermophilum*, pretreated rice straw has been converted to ethanol (8). Without removal of ethanol as a vapor, only 23 g/L could be produced owing to ethanol inhibiting the saccharification. With repeated flash cycling and additional solid substrate additions, up to 820 g of straw per L were converted at very high yield to ethanol, with condensate concentrations of 98–136 g/L of ethanol.

VIII. SUMMARY

Vacuum fermentation has been shown to dramatically increase the productivity of ethanolic fermentations through the removal of ethanol inhibition. By removing ethanol as it is formed, much more highly concentrated feeds can be fermented. This reduces the size of all materials-handling equipment (due to less water added) and, if vacuum distillation is employed, reduces product separation costs. Carbohydrate feedstocks remain by far the largest contributor to the cost of ethanol, and thus vacuum fermentation can reduce the manufacturing costs by a relatively small amount (typically a few cents/L). The reduced amount of yeast by-product formed in these systems also reduces credits for these by-products. Flash fermentation may be quite appropriate for ethanol production by *Z. mobilis*. Although this has no by-product, it can provide higher yields from the carbohydrate feedstocks. Further economic analysis of this process is warranted.

REFERENCES

1. Aigar, A. S., and Leudeking, R., A kinetic study of the alcoholic fermentation of glucose by *Saccharomyces cerevisiae, Chem. Eng. Prog. Symp.* 62: 57–59 (1969).
2. Aiba, S., Shoda, M., and Nagatani, M., Kinetics of product inhibition in alcoholic fermentation, *Biotechnol. Bioeng.* 10:845–864 (1968).
3. Holzberg, I., Finn, R. K., and Steinkraus, K. H., A kinetic study of alcoholic fermentation of grape juice, *Biotechnol. Bioeng.* 9:413–427 (1967).
4. Bazua, C., Effect of alcohol concentration on the kinetics of ethanol production by *Saccharomyces cerevisiae*, M.S. Thesis, University of California, Berkeley (1975).
5. van Uden, N., Ethanol toxicity and ethanol tolerance in yeasts, *Annu. Rep. Ferment. Proc.* 8:11–58 (1985).
6. Maiorella, B. L., Wilke, C. R., and Blanch, H. W., *Adv. Biochem Eng.* 20: 1 (1981).
7. Emert, G. H., and Katzen, R., Gulf's cellulose-to-ethanol process, *Chemtech* 10:610–614 (1980).

8. Ghose, T. K., Roychoudhury, P. K., and Ghosh, P., Simultaneous saccharification and fermentation (SSF) of lignocellulosics to ethanol under vacuum cycling and step feeding, *Biotechnol. Bioeng.* 26:377-381 (1984).

9. Wang, D. I. C., Avgerinos, G. C., Biocic, I., Wang, S-D., and Fang, H. Y., Ethanol from cellulosic biomass, *Philos. Trans. Roy. Soc. Lond.* 300:323-333 (1983).

10. Kirschbaum, E., and Gerstner, F., *Verfahrenstechnik* 1:10 (1939).

11. Roychoudhury, P. K., Ghose, T. K., Ghosh, P., and Chotani, G. K., Vapor liquid equilibrium behavior of aqueous ethanol solution during vacuum-coupled simultaneous saccharification and fermentation, *Biotechnol. Bioeng.* 28:972-976 (1986).

12. Maiorella, B. L., Blanch, H. W., and Wilke, C. R., Feed component inhibition in ethanolic fermentation by *Saccharomyces cerevisiae, Biotechnol. Bioeng.* 26:1155-1166 (1984).

13. Maiorella, B. L., Blanch, H. W., and Wilke, C. R., By-product inhibition effects on ethanol fermentation by *Saccharomyces cerevisiae, Biotechnol. Bioeng.* 25:103-121 (1983).

14. Ramalingham, A., and Finn, R. K., The Vacuferm process: A new approach to fermentation alcohol, *Biotechnol. Bioeng.* 19:583-589 (1977).

15. Cysewski, G. R., and Wilke, C. R., Rapid ethanol fermentations using vacuum and cell recycle, *Biotechnol. Bioeng.* 19:1125-1143 (1977).

16. Cysewski, G. R., Fermentation kinetics and process economics for the production of ethanol, Ph.D. Thesis, University of California, Berkeley (1976).

17. Wilke, C. R., Maiorella, B. L., Blanch, H. W., and Cysewski, G., U.S. Patent 4,359,533 (1982).

18. Blanch, H. W., and Wilke, C. R., Process development studies on the bioconversion of cellulose and production of ethanol, Report to the Solar Energy Research Institute, May 1981.

19. Maiorella, B. L., Blanch, H. W., and Wilke, C. R., Economic evaluation of alternative ethanol fermentation processes, *Biotechnol. Bioeng.* 26:1003-1025 (1984).

20. Ghose, T. K., and Tyagi, R. D., Rapid ethanol fermentation of cellulose hydrolysate, *Biotechnol. Bioeng.* 21:1387-1400 (1978).

21. Maiorella, B. L., Blanch, H. W., and Wilke, C. R., Rapid ethanol production via fermentation, Presented at the AICHE meeting, San Francisco, November, 1979.

22. McCabe, W. L., and Smith, J. C., *Unit Operations of Chemical Engineering*, 3rd ed. McGraw-Hill, New York (1976).

23. Zeikus, J. G., Chemical and fuel production by anaerobic bacteria, *Annu. Rev. Microbiol.* 34:423-464 (1980).

24. Ng, T. K., Ben-Bassat, A., and Zeikus, J. G., Ethanol production by thermophilic bacteria: Fermentation of cellulosic substrates by cocultures of *Clostridium thermocellum* and *Clostridium thermohydrosulfuricum, Appl. Environ. Microbiol.* 41:1337-1343 (1981).

25. Zeikus, J. G., and Ng, T. K., Thermophilic saccharide fermentations, *Annu. Rep. Ferment. Proc.* 5:263-289 (1982).

26. Sundquist, J. A., Blanch, H. W., and Wilke, C. R., Ethanol production with *Clostridium thermohydrosulfuricum*, presented at the 192nd ACS meeting, Anaheim, California, September 1986.

27. Sundquist, J. A., Ethanol production by *Clostridium thermohydrosulfuricum*, Ph.D., Thesis, University of California, Berkeley (1987).

28. Lee, K. J., Skotnicki, M. L., Tribe, D. E., and Rogers, P. L., *Biotechnol. Lett.* 2:339–344 (1980).

29. Lee, J. H., Woodard, J. C., Pagan, R. J., and Rogers, P. L., Vacuum fermentation for ethanol production using strains of *Zymomonas mobilis, Biotechnol. Lett.* 3:177–182 (1981).

11

Bioconversions in Permeabilized Cells

Hansruedi Felix

AGRO Research, Sandoz Agro Ltd.
Basel, Switzerland

I. INTRODUCTION

Whole cells made permeable by chemical or physical treatment allow the conversion of substrates to useful, expensive products. Permeabilized cells are useful if substrate or product or both cannot enter the cell. As many commercially interesting bioconversions require cofactor-dependent enzymes, regeneration of expensive cofactors is necessary. This is possible in permeabilized cells. Thus, permeabilized cells are useful in bioconversions in which cofactors are required and substrates or products or both are not taken up or released by intact cells.

The plasma membrane and the cell wall protect the living plasma from the environment. They create and preserve good conditions for cell metabolism. They play an important role in the regulation of substrate exchange between the cell and the surrounding medium. By active transport, certain substances can be accumulated within the cell. On the other hand, the penetration of molecules can be limited or totally prevented; for example, phosphorylated compounds cannot enter the intact cell. A series of methods has been developed to partially or totally overcome this barrier (1). Cells can be permeabilized without lysis of cells or destruction of enzyme systems. The morphology of the cells remains intact; low-molecular-weight molecules can freely enter and leave the cell. After permeabilization with organic solvents the cells are most often no longer viable, which can be useful, as energy is no longer wasted for the synthesis of cell mass.

The viability depends largely on the concentration of the agent used and the time of its application.

Enzymes are capable of catalyzing complex and specific reactions efficiently and under mild conditions. Of the enzymes currently exploited industrially, almost all are hydrolases because of their ready availability at low cost and their relative simplicity of action. They do not require low-molecular-mass cofactors or coenzymes for catalytic activity. Cofactors and coenzymes are expensive, and thus regeneration is a necessity. Regeneration is possible in permeabilized cells as the necessary substrates can freely enter the cells. Often regeneration and bioconversion are carried out by the same organism.

It is a matter of specific circumstances whether treatment with organic solvents is defined as permeabilization or extraction. Publications are mentioned here that discuss the use of organic solvents to permeabilize membranes. In some cases solubilization of substrates or products with organic solvents can be combined with membrane permeabilization. Product removal from a reactor (extractive bioconversions) is not considered here (2). In the future the integration of bioconversion, permeabilization, and downstream processing will certainly open new possibilities.

II. PERMEABILIZATION OF CELLS

For bioconversions, cells (microorganisms and plants) are very often immobilized. External and internal mass-transfer limitations influence the rate of substrate conversion by immobilized biocatalysts in artificial and natural systems (3). For a given immobilized biocatalyst the effectiveness of a conversion can be increased to the theoretical maximum by increasing the external mass-transfer coefficient and/or the permeability coefficient. The latter is important in the context of this chapter and can be increased by treating cells with compounds that change membrane permeability (1). Depending on the membrane permeability characteristics of educt and product of an enzymatic conversion, permeabilization is necessary.

For bioconversions, microorganisms were permeabilized with toluene (4–7), benzene + n-heptane (8), benzene + n-hexane + sorbitane monolaurate (9), ethanol (10), and a bile extract (11). Microorganisms can also be rendered permeable by different ways of drying: air drying (6,12,13), freeze drying (14), and acetone drying (5,6,8). Sonication combined with an activation at $37°C$ (37) or pH treatment for activation (15,16) also makes cells permeable. The choice of a special permeabilization method depends on the composition of the cell wall and the cell membrane. Among other things the difference between gram-positive and gram-negative bacteria must be taken into account.

Plant cells are most frequently treated with dimethylsulfoxide (DMSO), known to extract sterols from eukaryotic membranes. In some bioconversions a reversible permeabilization would be interesting. Several methods exist, such as phenethyl alcohol treatment of microorganisms (17,18) and DMSO treatment of plant cells (19). Plant cells entrapped in agarose or alginate were permeabilized with DMSO and cell viability was retained. A specific concentration of DMSO and application time is necessary. *Catharanthus roseus* cells intermittently treated with DMSO produced the ajmalicine isomers. DMSO also permeabilizes cells of the plant *Cinchona ledgeriana* and allows the release of alkaloids (20). However, the cells appear to be permanently damaged by the levels of DMSO necessary to produce release. In this respect DMSO is unsuitable for the long-term harvesting of alkaloids from these *C. ledgeriana* cells. Clearly, the suitability of DMSO for permeabilization varies with the culture. Part of the reason for this may be related to the conditions necessary for product release. In *C. roseus* a 30 min treatment with 5% DMSO produced nearly complete release of alkaloid (19). In *C. ledgeriana* such a treatment was ineffective and much higher DMSO levels were necessary. This may reflect the heterogeneity of the cell population but may also be due to the alkaloids being stored primarily in the vacuole, which requires slightly higher levels of DMSO for permeabiliziation than does the plasmalemma (21). In the case of *C. roseus* there may have been a rapid exhange of alkaloid between the vacuole and the cytoplasm that allowed a substantial release of alkaloid following permeabilization of the plasmalemma alone. Other possible explanations for the differences in behavior between *C. roseus* and *C. ledgeriana* can also be proposed. It has been shown, for example, that there is considerable cell-to-cell variation in the accumulation of alkaloids in *C. roseus* cultures (22) and possibly those cells in which alkaloid is concentrated are particularly sensitive to DMSO. This may thus allow extensive release of alka-loids with little effect on overall viability in this species. It must also be men-tioned that *C. roseus* cells were immobilized (19), whereas *C. ledgeriana* (20) were not. Whatever the exact explanation for the different behavior of different species, it is apparent that the DMSO permeabilization method is not universally applicable. The choice of an appropriate method depends on the organism and the composition of the cell wall and cell membrane. The kind of enzymatic reac-tion to be investigated is also important (1).

Immobilized cells exist in an environment of reduced water activity resulting in distinct alterations in metabolic behavior. In combination with limited oxygen supply, some mysteries reported for immobilized cells can be explained (23). The immobilization itself can lead to altered membrane permeability. After im-mobilization in alginate, *Mucuna pruriens* cells release 90% of the product dihy-droxyphenylalanine (24). A controlled release of oxalate from *Amaranthus tri-color* after immobilization and permeabilization with chitosan was observed (25).

A simple method, employing high-voltage electrical discharge (electropora-tion), was developed to introduce phosphorylated nucleosides into the cyto-plasm of viable cells (26). The same method has been applied for the release of secondary metabolites from plant cells (27). It is, however, not clear whether this method can also be used for large reactors.

III. REGENERATION OF COFACTORS OR COENZYMES

Of the about 2000 different enzymes to which a specific number has been assigned, about one-third require one of the four adenine coenzymes—NAD, NADP, CoA, and ATP—to participate in enzymtic reactions. Many of the enzymes likely to be useful for the production of commodity chemicals and fine chemicals require cofactors or an energy-rich cosubstrate; without an efficient regeneration system any such process is rendered impossible for economic reasons. Of those enzymes exploited industrially almost all are hydrolases because of their ready availability at low cost and their relative simplicity of action.

There are several methods of recycling oxidized and reduced forms of NAD(P) (28): chemically (autooxidizable electron acceptors), electrochemically (electrodes), enzymatically (isolated enzymes, and permeabilized cells), and photochemically. The enzymatic regeneration is preferred at present. Immobilized, permeabilized cells of the plant *C. roseus* reduce the alkaloid cathenamine to ajmalicine. The NADPH required in the transformation is regenerated by isocitrate through the isocitrate dehydrogenase found in the same cell (Fig. 1) (29). By chemical stabilization methods these systems were shown to be active for long periods, up to 2 months (30). To allow their use in flow-through systems immobilization of NAD(P)$^+$ to water-soluble polyacrylamide derivatives of molecular mass of approximately 5–10 kD can be attempted (28,31). These preparations can enter various permeabilized cells, leading to efficient regeneration. An interesting method of NADP$^+$ reduction was found in the system of a methanogen using formic acid or hydrogen as electron donor (32). Work must be carried out under anaerobic conditions, which is certainly a disadvantage of the method, which is not outweighed by the advantage of cheap precursors.

Extensive studies have been carried out on bioreactors, combining ATP-consuming and ATP-generating systems with immobilized biocatalysts to produce complex compounds in microorganisms (see Figs. 2 through 5 and Table 1). The study of the synthesis of gramicidin S conjugated with ATP-regenerating system using acetate kinase with acetylphosphate as substrate is well known. However, from the viewpoint of practical use, enzyme activities and their operational stabilities in such systems are sometimes unequal, and it is very difficult to maintain whole enzyme systems in a stable state for long periods of time. Moreover, supplies of acetylphosphate at a low price cannot be expected because its mass production has not been attempted. Nevertheless, acetate kinase using acetylphosphate to regenerate ATP has been used in different bioreactor systems

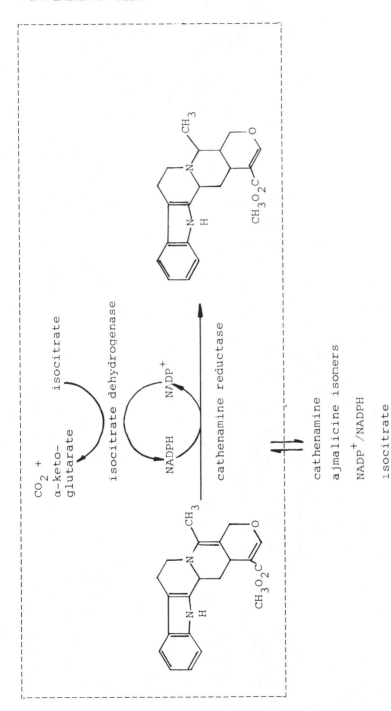

Figure 1 Conversion of cathenamine to ajmalicine isomers (ajmalicine, 19-epi-ajmalicine, and tetrahydroalstonine) using permeabilized cells of the plant *C. roseus* and NADPH as coenzyme. NADPH is built up with isocitrate dehydrogenase from NADP⁺: (− − −) cell wall with permeable cell membrane: (⇌) free diffusion of substrates, products, and cofactors. (From Ref. 29.)

Table 1 Production of Useful Compounds in Permeabilized, Immobilized Microorganisms

Product	Educt	Cofactor regeneration	Productivity	References
Glutathione	glu, cys, gly	Glycolysis or acetate kinase	16.6[a]	4[b], 5–7, 10, 33, 34
NADP	NAD	Glycolysis	1.1[c]	5, 6, 36[b]
S-adenosylmethionine	met, adenosine	Glycolysis	0.1[a]	6[b]
CDP choline	CMP, choline	Glycolysis	2[d]	9[b], 12, 14
Coenzyme A	Pantothenic acid, cys, AMP	Glycolysis	0.2[d]	9[b], 13
L-alanine	Ammonium fumarate	—	3,000[a]	15, 16[b]
L-alanine	L-aspartic acid	—	8,000[c]	37[b]
L-malic acid	Fumarate	—	10,000[c]	11[b]
Prednisolone	Hydrocortisone	—	—	8, 39–42
Androst-1,4-diene 3,17-dione	4-Androstene-3,17-dione	—	—	8
Δ[1]-Cortexolone	Cortexolone	—	—	38
3-Keto-Δ[4]-steroids	β-Sitosterol, stigmasterol	—	—	44
Cortisol	Cortexolone	—	—	43
Cholestenone	Cholesterol	—	29[c]	44, 45[b]
Progesterone	Pregnenolone	—	—	45
4-Androstene-3,17-dione	Dehydroepiandrosterone	—	—	44, 45
1-Menthol	d,l-Menthyl succinate	—	—	49

[a]Expressed as μmol/h·ml gel.
[b]Values taken from references. These values are not easy to compare, as different immobilization methods were used.
[c]Expressed as μmol/h·g cells.
[d]Expressed as mmol/h.

(4,6,7). Immobilized microbial cells can be utilized as a tool to regenerate ATP. Yeast cells phosphorylate adenosine, AMP, and ADP to yield ATP using glycolytic energy (5,6,9,10,12,14,33,34). A stable ATP-generating system was prepared by entrapping *Saccharomyces cerevisiae* cells in polyacrylamide gel (5). The half-life of the column was 19 days at a space velocity of 0.3 h^{-1} at 30°C, and ATP was regenerated at 7 μmol/h and ml of gel.

Dextran-bound ADP was also used (6,7). However, the glutathione-producing activity of these gels was rather low, probably for the following reasons. First, the cosubstrate activity of dextran-bound ATP is lower than that of ATP itself. Second, steric interferences restricted the availability of ATP. Third, in the reaction system employed, the dextran-bound ATP is involved in the three sequential enzyme reactions catalyzed by acetate kinase and the constituents of the glutathione-forming enzyme system. Probably a single form of dextran-bound ATP does not show high activity in all three enzyme reactions.

ATP regeneration by immobilization of thylakoids and chromatophores for hydrogen production represents a rather unconventional method. High-added-value compounds are produced by direct conversion of solar energy in immobilized permeabilized (sonicated) *Rhodopseudomonas capsulata* (35).

IV. APPLICATION OF PERMEABILIZED CELLS

Although permeabilized cells free in solution show higher enzymatic activities than do immobilized permeabilized cells (4), only the latter were investigated to any great extent, as their operational stability is much larger. Bioconversions were carried out in microorganisms and plant cells. Industrial application seems possible.

A. Microorganisms

The ATP-generating systems mentioned were combined with ATP-consuming systems in the same microbial cells or in other cells (Table 1).

When CMP and choline were added to the ATP-generating system (glycolysis) of *S. cerevisiae*, CDP choline formation was observed (Fig. 2) (9,12,14).

S. cerevisiae cells entrapped in polyacrylamide gel were used for the continuous production of glutathione from L-glutamate, L-cysteine, and glycine using the glycolytic system in the same organism as energy source (Fig. 3) (33,34). For continuous operation, a supply of a small amount of NAD$^+$ was necessary, as in the case of ATP and CDP choline production. The column reactor was stable, and its half-life was about 23 days. Because of its simplicity this system of *S. cerevisiae* (33,34) was found to be superior to the coimmobilized system of *Escherichia coli* (synthesizing glutathione) and *S. cerevisiae* (generating ATP) (5,6). The highest activity was found in an immobilized *E. coli* system using

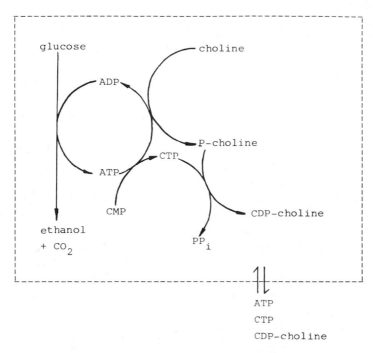

ATP
CTP
CDP-choline

Figure 2 CDP choline is formed by dried immobilized yeast cells representing a kind of bioreactor. The transformation of choline to CDP choline and the regeneration of ATP is carried out by the same organism. Symbols are explained in Fig. 1. (From Ref. 12.)

acetylphosphate as phosphate donor for ATP generation (4). As already mentioned, a trial using dextran-bound ATP as a cofactor was made (6,7). The glutathione-producing activity of these gels, however, was rather low.

In general there are problems with glutathione production in immobilized, permeabilized cells: (1) feedback inhibition of γ-glutamylcysteine synthetase by glutathione; (2) inhibition of γ-glutamylcysteine synthetase by ADP; (3) glutathione degrading activity; (4) glutathione transport barrier across the cell membrane; and (5) inhibition of glutathione synthesis by phosphate. Some of these have already been overcome, but the inhibitory effect of phosphate on glutathione synthesis seems to be the barrier to further development of the glutathione production system utilizing the glycolytic pathway as a tool for ATP generation. No such problems occur with *Candida tropicalis*, which produces extracellular glutathione during growth in filamentous form caused by adding 2.5% (v/v) ethanol (10). Myoinositol added at a physiological concentration

(5 μg/ml) prevented the ethanol-induced production of extracellular glutathione, probably by stabilizing membranes. It is probable that the release of glutathione to the culture medium is a consequence of the effects of ethanol on the structure and function of cell membranes, especially the plasma membrane.

A coimmobilized system was also constructed for the production of CoA from pantothenate and cysteine (Fig. 4). *Brevibacterium ammoniagenes* (synthesizing CoA) and *S. cerevisiae* (generating ATP) showed a higher activity of CoA synthesis using glucose as energy source than immobilized *B. ammoniagenes* alone supplemented with ATP (9,13).

Nicotinamde adenine dinucleotide phosphate ($NADP^+$) was produced by the phosphorylation of NAD^+ with *B. ammoniagenes* having NAD kinase activity and *S. cerevisiae* having glycolytic activity (Fig. 5). The system was stable over 10 days, maintaining the activity of $NADP^+$ formation of about 0.6 μmol/h and ml gel at a space velocity of 0.3 h^{-1} at 30°C (5,6).

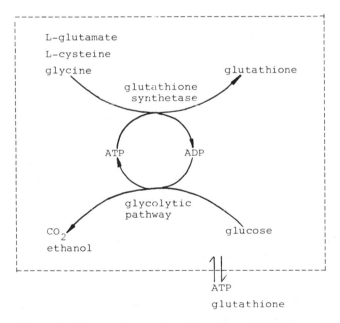

Figure 3 The reactor for glutathione contains two reaction systems. One is the ATP-regenerating system containing the enzymes involved in the glycolytic pathway. By this system ADP is converted to ATP, consuming glucose. This system also allows the regeneration of NAD^+ in the course of the degradation of glucose to ethanol. The other is the glutathione synthetase I and II system. Yeast cells were used as bioreactor. The cells became permeable to glutathione after immobilization in polyacrylamide gel. (From Refs. 33 and 35.)

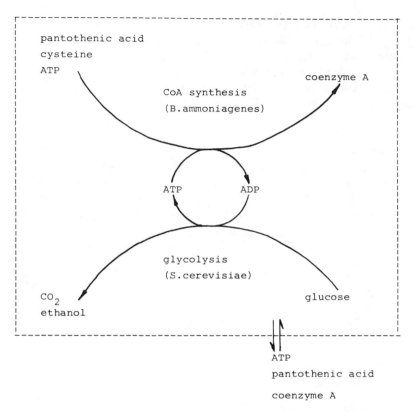

Figure 4 Synthesis of coenzyme A in a reactor system in which *S. cerevisiae* and *B. ammoniagenes* are coimmobilized. Both cell systems were permeabilized by *n*-hexane. *S. cerevisiae* produced ATP in the course of glycolysis, and *B. ammoniagenes* carried out the coenzyme A synthesis. (From Refs. 9 and 13.)

ATP regenerated by the glycolytic pathway in acetone-dried *S. cerevisiae* can also be supplied to air-dried *S. cerevisiae* producing S-adenosyl-L-methionine (6). It was produced at a rate of 0.1 μmol/h per ml gel.

A series of other reactions was performed in permeabilized cells: L-malic acid synthesis from potassium fumarate in *Brevibacterium flavum* (11), L-alanine synthesis from L-aspartic acid using pyridoxal phosphate as cofactor in this decarboxylation reaction (37), and L-alanine from ammonium fumarate (15,16) in *Pseudomonas dacunhae*.

Enzymatic and fermentative bioconversions, even when water-insoluble, lipophilic compounds are substrates, have usually been carried out in aqueous systems because enzymes and microbial cells were believed to be unstable in organic solvents. Immobilization often provides biocatalyst resistance against denaturation by organic solvents. Thus, conversion of water-insoluble compounds, such as steroids, have been carried out with immobilized microbial cells in organic solvent systems. The hydrophobicity of gels entrapping microbial cells sometimes seriously affects the efficiency of such conversions. Water-miscible organic solvent homogeneous systems, water-immiscible organic solvent two-phase systems, and organic solvent systems can be used. Water-miscible organic solvents, such as methanol, are often useful in constructing homogeneous reaction mixtures with immobilized microbial cells. Dehydrogenation of Reichstein's compound S by *Pseudomonas testosteroni* (38) and of hydrocortisone by *Arthrobacter simplex* (39,40) was carried out in aqueous systems containing 10% methanol. Methanol was also utilized for the bioconversion of hydrocortisone (41,42) and of Reichstein's compound S (43) by immobilized growing cells.

Acetone-dried *Arthrobacter simplex* and benzene + *n*-heptane–treated *Nocar--dia rhodocrous* were used for the conversion of hydrocortisone to prednisolone and 4-androstene-3,17-dione to androst-1,4-diene-3,17-dione, respectively (8).

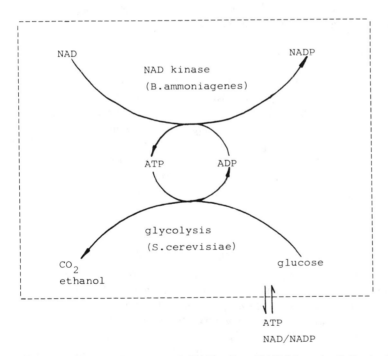

Figure 5 Phosphorylation of NAD using NAD kinase activity in *B. ammonia-genes* and glycolysis in *S. cerevisiae* as the ATP-generating system. Symbols are explained in Fig. 1.

A few reports have been published about the application of water-organic solvent two-phase systems and organic solvent systems. Free microbial cells seem to be unstable in the latter system, and the former systems are not applicable to homogeneous and continuous bioconversion systems. Immobilized *N. rhodocrous* cells catalyzed the dehydrogenation of steroids at several positions, even in organic solvents (44,45). Problems can arise if hydrophobic or hydrophilic gels are used in more or less hydrophilic solvents or solvent mixtures (46–48). Stereoselective hydrolysis of *d,l*-menthyl succinate to yield *l*-menthol has been carried out in water-saturated *n*-heptane by *Rhodotorula minuta* (49). The entrapped cells had a half-life of 63 days, whereas the half-life of the free cells was only 2 days.

Although many organic compounds have been excluded from enzymatic reactions by their low solubility in aqueous solution, the examples mentioned here may encourage transformations of various hydrophobic compounds by using immobilized cells in organic solvents, as it has been common practice with isolated enzymes for quite some time (e.g., Ref. 50). One limitation of these methods exists. Enzymes dependent on membrane systems, such as mixed-function oxygenases carrying out hydroxylations, are excluded because organic solvents destroy the essential membrane systems.

B. Plant Cells

In plant cells the advantageous utilization of immobilized biocatalysts may be impeded because the produced metabolites of interest are often stored within the cell. The permeability of the cell membrane can be changed upon immobilization, either as a side effect of the immobilization method or by design. Products that can move within a plant across membranes either by diffusion or by an active transport mechanism are likely to be exported from cells in culture, at least under certain conditions. If a particular culture does not release such a product, then altering the culture conditions may induce the cells to develop the relevant transport mechanism.

Immobilization affects the partitioning of products between cells and the culture medium. In *C. roseus*, free cells retain all of their alkaloid products within the vacuole but immobilized cells release up to 90% of the alkaloid produced into the medium. This effect on release of the product occurs in cells that are immobilized either in alginate (51) or in polyacrylamide-xanthan gum (52). In cells of *M. pruriens* immobilization in alginate leads to the release into the medium of L-dihydroxyphenylalanine (L-DOPA) formed from exogenously supplied L-tyrosine; free cells of the same line make L-DOPA by de novo synthesis and store the product intracellularly (24). The modified metabolic behavior of immobilized cells compared with free cells can be traced to the chemical and physical properties of the support material and the effect this has on the microenvironment (23). Charged groups on the polymer bind water and so effectively reduce the water activity in the vicinity of cells in the gel, which leads to stress, increased maintenance metabolism, and reduced growth. The different factors can vary from one support matrix to another.

In conditions in which the vacuolar pH is lower than that of the surrounding medium, alkaloids are trapped in the vacuole in the protonated form; this lowers the internal concentration of the uncharged alkaloid and, consequently, inward diffusion. This proposition was tested with tabernanthine in *C. roseus* and *Acer pseudoplatanus* (53).

Membrane properties can also be manipulated externally. Table 2 lists some of the plants permeabilized and the bioconversion carried out. As mentioned earlier, an intermittent release of alkaloids from *C. roseus* is possible (19).

Treatment of other plants with the same method resulted in cell death (20, 54,55). Relatively high DMSO concentrations were necessary to release intracellularly stored quinoline alkaloids from *C. ledgeriana* (20). Furthermore, alginate-entrapped cells of *M. pruriens* hydroxylating tyrosine to DOPA were premeabilized by treatment with isopropanol (55). The hydroxylation activity of the immobilized cells was improved threefold by this treatment, but already in a second batch most of the enzymatic activity was lost owing to leakage of the enzyme from the permeabilized cells. There is a relationship between the concentration of permeabilizing agent and the degree of permeabilization. Various cell lines respond differently. Only tolerant cell lines may be permeabilized with preserved viability.

In several known methods of recovering chemical compounds from cells, a membrane-modifying agent like chloroform is used to promote the extraction of the chemical compounds. The use of such agents gives rise to substantial cell damage and in many circumstances destruction of the cells. For example, the use of chloroform can be shown to cause massive lysis of the cells. It has been found that, surprisingly, cells can be induced to excrete intracellularly produced chemical compounds if they are brought together and maintained in close proximity to a sufficient extent, that is, if they are maintained in a sufficiently high cell density (52).

As in microorganisms, the problem of cofactor recycling exists in plants. The reduction of cathenamine to ajmalicine isomers requires NADPH as cofactor (29). The most efficient transformation was observed when cathenamine together with NADPH was added to DMSO-treated immobilized cells. The NADPH level can be held at the initial concentration if isocitric acid is added (Fig. 1). A somewhat lower transformation rate was observed on replacement of the reduced coenzyme by $NADP^+$ and isocitric acid.

Permeabilized cells do not carry out reactions that are catalyzed by membrane-dependent enzymes. In DMSO-treated *Haplopappus gracilis* cells only the conversion of naringenin to eriodictyol and dihydrokaempferol could be

Table 2 Bioconversions in Permeabilized Plant Cells

Plant species	Bioconversion	Immobilization method	Permeabilization method	References
Catharanthus roseus	Cathenamine to ajmalicine	Agarose	Ether, DMSO, cytochrome c, polylysine, protamine sulfate, chloroform	29, 30, 51, 60
Nicotania tabacum	D-tryptophan to L-tryptophan	—	DMSO	61
Haplopappus gracilis	Naringenin to eriodictyol and dihydrokaempferol	—	DMSO	56
C. roseus	Tryptamine and secologanin to cathenamine and ajmalicine	Agarose	DMSO	19
Thalictrum rugosum	Sucrose to berberine	—	DMSO, phenethyl-alcohol, chloroform, Triton X-100, hexadecyl-trimethylammonium bromide (HDTMAB)	54
Cinchona ledgeriana	Sucrose to cinchonine and cinchonidine	—	DMSO	20
Mucuna pruriens	Tyrosine to DOPA	Alginate	Isopropanol	55
Amaranthus tricolor, Asclepias syriaca	Sucrose to oxalate and proteins	Chitosan	Chitosan	62
C. roseus	Sucrose to ajmalicine and	Acryl-amide gel	Chloroform	52

observed; the subsequent hydroxylation was only found with a microsomal preparation (56).

Except for certain specialist biotransformations, it seems unlikely that true immobilized plant cells can compete effectively in this area with a less complex biocatalyst, such as immobilized enzymes or microbial cells.

C. Industrial Application

With the exception of living cells, many applictions of immobilized biocatalyst systems are limited by the high cost of coenzymes or the lack of a suitable coenzyme regeneration mechanism. As mentioned, the solution of several problems was tried using permeabilized cells. Thus glutatione, CDP choline, coenzyme A, $NADP^+$ and S-adenosylmethionine were synthesized by microbial cells that simultaneously supplied the necessary cofactors. The remarkably high yields obtained suggest potential application for production on a larger scale (Table 1).

V. IMMOBILIZATION

Immobilization must have the following features (8). (1) Microbial or plant cells must be easy to entrap in gel matrices without loss of the enzymes or cells. (2) Extended kinds of prepolymers should be applicable to the immobilization of biological materials. Immobilization has thus far been restricted to hydrophilic carriers. (3) In the case of transformations of water-insoluble substrates, such as steroids, terpenoids, and various organic compounds, hydrophobic gels are considered superior to hydrophilic gels as far as the diffusion characteristics of these substrates are concerned. The transformation of organic compounds by immobilized cells and enzymes in organic solvents, in which the concentration of lipophilic substrates can be increased, is very attractive since high product yields are favored. In such systems, hydrophobic gel-entrapped cells are expected to be more efficient catalysts than hydrophilic gel-entrapped cells.

Permeabilized C. roseus cells immobilized in alginate did not carry out any reaction dependent on phosphorylated cofactors (29). This is most likely due to Ca^{2+} present in the incubation mixture (required for stabilization of the alginate gel), which prevents the reaction by ionic binding to the negatively charged phosphorylated coenzymes. The enzymatic activities can, however, be regained by dissolving the immobilized plant cells through addition of a Ca^{2+}-complexing agent, such as citric acid. Thus for certain reactions a neutral carrier, such as agarose or a polyurethane foam, is adequate.

VI. PROBLEMS

Using permeabilized, immobilized cells for bioconversion, the problems that arise are similar to those with immobilized enzymes or intact cells: CO_2 evolution can destroy the carrier (37), side reactions occur, and stability is inadequate.

In several cases side reactions could effectively be suppressed. After treatment with a bile extract fumaric acid was preferentially converted to malic acid; succinic acid was only a minor product (11). pH treatment is used for the physical deactivation of unwanted enzymatic reactions (15,16,57-59).

VII. CONCLUSIONS

Permeabilized cells can be a good tool for bioconversions, especially in microorganisms. As permeabilization of cells is most often achieved with organic solvents, the latter can serve at the same time for the solubilization of hydrophobic substrates and products. In many cases authors showed the feasibility of using permeabilized cells.

REFERENCES

1. Felix, H., Permeabilized cells, *Anal. Biochem.* 120:211-234 (1982).
2. Mattiasson, B., and Larsson, M., Extractive bioconversions with emphasis on solvent production, *Biotechnol. Genet. Eng. Rev.* 3:137-174 (1985).
3. Kasche, V., Correlation of experimental and theoretical data for artificial and natural systems with immobilyzed biocatalysts, *Enzyme Microb. Technol.* 5:2-13 (1983).
4. Murata, K., Tani, K., Kato, J., and Chibata, I., Glutathione production coupled with an ATP regeneration system, *Eur. J. Appl. Microbiol. Biotechnol.* 10:11-21 (1980).
5. Murata, K., Tani, K., Kato, J., and Chibata, I., Glycolytic pathway as an ATP generation system and its application to the production of glutathione and NADP, *Enzyme Microb. Technol.* 3:233-242 (1981).
6. Murata, K., Tani, K., Kato, J., and Chibata, I., Continuous production of glutathione using immobilized microbial cells containing ATP generation system, *Biochimie* 62:347-352 (1980).
7. Murata, K., Tani, K., Kato, J., and Chibata, I., Application of immobilized ATP in the production of glutathione by a multienzyme system, *J. Appl. Biochem.* 1:283-290 (1979).
8. Omata, T., Tanaka, A., Yamane, T., and Fukui, S., Immobilization of microbial cells and enzymes with hydrophobic photo-crosslinkable resin prepolymers, *Eur. J. Appl. Microbiol. Biotechnol.* 6:207-215 (1979).
9. Samajima, H., Kimura, K., Ado, Y., Suzuki, Y., and Tadokoro, T., Regeneration of ATP by immobilized microbial cells and its utilization for the synthesis of nucleotides, *Enzyme Eng.* 4:237-244 (1978).

10. Yamada, Y., Tani, Y., and Kamihara, T., Production of extracellular glutathione by *Candida tropicalis Pk 233*, *J. Gen. Microbiol.* 130:3275–3278 (1984).

11. Takata, I., Yamamoto, K., Tosa, T., and Chibata, I., Immobilization of *Brevibacterium flavum* with carrageenan and its application for continuous production of L-malic acid, *Enzyme Microb. Technol.* 2:30–36 (1980).

12. Kimura, A., Tatsutomi, Y., Mizushima, N., Tanaka, A., Matsuno, R., and Fukuda, H., Immobilization of glycolysis system of yeasts and production of cytidine diphosphate choline, *Eur. J. Appl. Microbiol. Biotechnol.* 5:13–16 (1978).

13. Ogata, K., Shimizu, S., and Tani, Y., Studies on the metabolism of pantothenic acid in microorganisms. Part I. Distribution of CoA-accumulating activity in microorganisms and isolation of reaction products, *Agric. Biol. Chem.* 36:84–92 (1971).

14. Ado, Y., Suzuki, Y., Tadokoro, T., Kimura, K., and Samejima, H., Regeneration of ATP by immobilized microbial cells and its utilization for the synthesis of ATP and CDP-choline, *J. Solid-Phase Biochem.* 4:43–55 (1979).

15. Takamatsu, S., Tosa, T., and Chibata, I., Production of L-alanine from ammonium fumarate using two microbial cells immobilized with κ-carrageenan, *J. Chem. Eng. Japan* 18:66–70 (1985).

16. Takamatsu, S., Tosa, T., and Chibata, I., Industrial production of L-alanine from ammonium fumarate using immobilized microbial cells of two kinds, *J. Chem. Eng. Japan* 19:31–36 (1986).

17. Treick, R. W., and Konetzka, W. A., Physiological state of *Escherichia coli* and the inhibition of deoxyribonucleic acid synthesis by phenethyl alcohol, *J. Bacteriol.* 88:1580–1584 (1964).

18. Silver, S., and Wendt, L., Mechanism of action of phenethyl alcohol: Breakdown of the cellular permeability barrier, *J. Bacteriol.* 93:560–566 (1967).

19. Brodelius, P., and Nilsson, K., Permeabilization of immobilized plant cells, resulting in release of intracellularly stored products with preserved cell viability, *Eur. J. Appl. Microbiol. Biotechnol.* 17:275–280 (1983).

20. Parr, A. J., Robins, R. J., and Rhodes, M. J. C., Permeabilization of *Cinchona ledgeriana* cells by dimethylsulphoxide. Effects on alkaloid release and long-term membrane integrity, *Plant Cell Rep.* 3:262–265 (1984).

21. Delmer, D. P., Dimethylsulfoxide as a potential tool for analysis of compartmentation in living plant cells, *Plant Physiol.* 64:623–629 (1979).

22. Deus, B., and Zenk, M. H., Exploitation of plant cells for the production of natural compounds, *Biotechnol. Bioeng.* 24:1965–1974 (1982).

23. Mattiasson, B., and Hahn-Hägerdal, B., Microenvironmental effects on metabolic behaviour of immobilized cells. A hypothesis, *Eur. J. Appl. Microbiol. Biotechnol.* 16:52–55 (1982).

24. Wichers, H. J., Malingré, T. M., and Huizing, H. J., The effect of some environmental factors on the production of L–DOPA by alginate-entrapped cells of *Mucuna pruriens, Planta* 158:482–486 (1983).

25. Knorr, D., and Teutonico, R. A., Chitosan immobilization and permeabilization of *Amaranthus tricolor* cells, *J. Agric. Food Chem.* 34:96–97 (1986).

26. Sokoloski, J. A., Jastreboff, M. M., Bertino, J. R., Sartorelli, A. C., and Narayanan, R., Introduction of deoxyribonucleoside triphosphates into intact cells by electroporation, *Anal. Biochem.* 158:272–277 (1986).

27. Brodelius, P., Shillito, R., and Potrykus, I., Verfahren zur Gewinnung von Pflanzeninhaltstoffen, Swiss Patent Application 04063/86-3 (1986).

28. Lowe, C. R., The application of coenzyme-dependent enzymes in biotechnology, *Philos. Trans. Roy. Soc. Lond.* B300:335–353 (1983).

29. Felix, H., Brodelius, P., and Mosbach, K., Enzyme activities of the primary and secondary metabolism of simultaneously permeabilized and immobilized plant cells, *Anal. Biochem.* 116:462–470 (1981).

30. Felix, H. R., and Mosbach, K., Enhanced stability of enzymes in permeabilized and immobilized cells, *Biotechnol. Lett.* 4:181–186 (1982).

31. Mosbach, K., The potential in biotechnology of immobilized cells and of immobilized multistep enzyme-coenzyme systems, *Philos. Trans. Roy. Soc. Lond.* B300:355–367 (1983).

32. Eguchi, S. Y., Nakata, H., Nishio, N., and Nagai, S., $NADP^+$ reduction by a methanogen using HCOOH or H_2 as electron donor, *Appl. Microbiol. Biotechnol.* 20:213–217 (1984).

33. Murata, K., Tani, K., Kato, J., and Chibata, I., Continuous production of glutathione by immobilized *Saccharomyces cerevisiae* cells, *Eur. J. Appl. Microbiol. Biotechnol.* 6:23–27 (1978).

34. Murata, K., Tani, K., Kato, J., and Chibata, I., Glutathione production by immobilized *Saccharomyces cerevisiae* cells containing an ATP regeneration system, *Eur. J. Appl. Microbiol. Biotechnol.* 11:72–77 (1981).

35. Cocquempot, M. F., Larreta Garde, V., and Thomas, D., Stabilization of biological photosystems: Immobilization of thylakoids and chromatophores for hydrogen production and ATP regeneration, *Biochimie* 62:615–621 (1980).

36. Ado, Y., Kimura, K., and Samejima, H., Production of useful nucleotides with immobilized microbial cells, *Enzyme Eng.* 5:295–304 (1980).

37. Yamamoto, K., Tosa, T., and Chibata, I., Continuous production of L-alanine using *Pseudomonas dacunhae* immobilized with carrageenan, *Biotechnol. Bioeng.* 22:2045–2054 (1980).

38. Yang, H. S., and Studebaker, J. F., Continuous dehydrogenation of a steroid with immobilized microbial cells: Effect of an exogenous electron acceptor, *Biotechnol. Bioeng.* 20:17–25 (1978).

39. Sonomoto, K., Jin, I-N., Tanaka, A., and Fukui, S., Application of urethane prepolymers to immobilization of biocatalysts: Δ^1-Dehydrogenation of hydrocortisone by *Arthrobacter simplex* cells entrapped with urethane prepolymers, *Agric. Biol. Chem.* 44:1119–1126 (1980).

40. Sonomoto, K., Tanaka, A., Omata, T., Yamane, T., and Fukui, S., Application of photo-crosslinkable resin prepolymers to entrap microbial cells. Effects of increased cell-entrapping gel hydrophobicity on the hydrocortisone Δ^1-dehydrogenation, *Eur. J. Appl. Microbiol. Biotechnol.* 6:325–334 (1979).

41. Larsson, P. O., Ohlson, S., and Mosbach, K., New approach to steroid conversion using activated immobilized microorganisms, *Nature* 263:796–797 (1976).
42. Ohlson, S., Larsson, P. O., and Mosbach, K., Steroid transformation by activated living immobilized *Arthrobacter simplex* cells, *Biotechnol. Bioeng.* 20:1267–1284 (1978).
43. Ohlson, S., Flygare, S., Larsson, P-O., and Mosbach, K., Steroid hydroxylation using immobilized spores of *Curvularia lunata* germinated in situ, *Eur. J. Appl. Microbiol. Biotechnol.* 10:1–9 (1980).
44. Omata, T., Iida, T., Tanaka, A., and Fukui, S., Transformation of steroids by gel-entrapped *Nocardia rhodocrous* cells in organic solvent, *Eur. J. Appl. Microbiol. Biotechnol.* 8:143–155 (1979).
45. Omata, T., Tanaka, A., and Fukui, S., Bioconversions under hydrophobic conditions: Effect of solvent polarity on steroid transformations by gel-entrapped *Nocardia rhodocrous* cells, *J. Ferment. Technol.* 58:339–343 (1980).
46. Fukui, S., and Tanaka, A., Immobilized microbial cells, *Annu. Rev. Microbiol.* 36:145–172 (1982).
47. Fukui, S., Ahmed, S. A., Omata, T., and Tanaka, A., Bioconversion of lipophilic compounds in non-aqueous solvent. Effect of gel hydrophobicity on diverse conversions of testosterone by gel-entrapped *Nocardia rhodocrous* cells, *Eur. J. Appl. Microbiol. Biotechnol.* 10:289–301 (1980).
48. Yamané, T., Nakatani, H., Sada, E., Omata, T., Tanaka, A., and Fukui, S., Steroid bioconversion in water-insoluble organic solvents: Δ^1-Dehydrogenation by free microbial cells and by cells entrapped in hydrophilic or lipophilic gels, *Biotechnol. Bioeng.* 21:2133–2145 (1979).
49. Omata, T., Iwamoto, N., Kimura, T., Tanaka, A., and Fukui, S., Steroselective hydrolysis of dl-menthyl succinate by gel-entrapped *Rhodotorula minuta* var. *texensis* cells in organic solvent, *Eur. J. Appl. Microbiol. Biotechnol.* 11:199–204 (1981).
50. Kazandjian, R. A., Dordick, J. S., and Klibanov, A. M., Enzymatic analyses in organic solvents, *Biotechnol. Bioeng.* 28:417–421 (1986).
51. Brodelius, P., Mosbach, K., Zenk, M., and Deus, B., Catalysts for the production and transformation of natural products having their origin in higher plants, process for production of the catalysts, and use thereof, Eur. Patent Appl. 0 022 434 A3 (1980)
52. Anthony, C., and Rosevear, A., Production of chemical compounds from viable cells, U.K. Patent Appl. GB 2 096 169 A (1982).
53. Renaudin, J-P., C-tabernanthine by cell suspension cultures of *Catharantus roseus* (L.) G.Don and *Acer pseudoplatanus* L., *Plant Sci. Lett.* 22:59–69 (1981).
54. Brodelius, P., Permeabilization of plant cells: Viability studies for release of intracellularly stored products, *Appl. Microbiol. Biotechnol.* 27:561–566 (1988).
55. Wichers, H. J., Malingré, T. M., and Huizing, H. J., Optimization of the

biotransformation of L-tyrosine into L-dihydroxyphenylalanine (DOPA) by alginate-entrapped cells of *Mucuna pruriens, Planta* 166:421–428 (1985).

56. Fritsch, H., and Grisebach, H., Biosynthesis of cyanidin in cell cultures of *Haplopappus gracilis, Phytochemistry* 14:2437–2442 (1975).

57. Umemura, I., Takamatsu, S., Sato, T., Tosa, T., and Chibata, I., Improvement of production of L-aspartic acid using immobilized microbial cells, *Appl. Microbiol. Biotechnol.* 20:291–295 (1984).

58. Tosa, T., Takamatsu, S., Furui, M., and Chibata, I., Continuous production of L-alanine: Successive enzyme reactions with two immobilized cells, *Ann. N.Y. Acad. Sci.* 434:450–453 (1984).

59. Chibata, I., Tosa, T., and Takamatsu, S., Industrial production of L-alanine using immobilized *Escherichia coli* and *Pseudomonas dacunhae, Microbiol. Sci.* 1:58–62 (1984).

60. Brodelius, P., Deus, B., Mosbach, K., and Zenk, M. H., Immobilized plant cells for the production and transformation of natural products, *FEBS Lett.* 103:93–97 (1979).

61. Miura, G. A., and Mills, S. E., The conversion of D-tryptophan to L-tryptophan in cell cultures of tobacco, *Plant Physiol.* 47:483–487 (1971).

62. Knorr, D., Miazga, S. M., and Teutonico, R. A., Immobilization and permeabilization of cultured plant cells, *Food Technol.* 39:135–142 (1985).

12

Ion-Pair Extraction in Biological Systems

Göran Schill

University of Uppsala
Uppsala, Sweden

Bengt-Arne Persson

AB Hässle
Mölndal, Sweden

I. INTRODUCTION

The extraction methods for organic compounds have developed rapidly during the last two decades owing to the need for efficient and mild isolation procedures for minor components in biomedical, biochemical, and environmental studies, and new principles for liquid-liquid and liquid-solid distribution have been exploited.

The problems of extractions from biological materials can be highly intricate. A minor component must often be separated from a series of organic compounds of a nature and a concentration that are known to a fairly limited extent. Isolation methods that aim at the removal of disturbing compounds are as a rule inefficient. Instead it is necessary to base the methodological work on the known properties of the compound of interest and to make a procedure that is as specific as possible. The general principles to achieve this goal are well known.

The differences in the distribution ratio between the sample components can be improved by using phase materials with selective interactions with the sample components. This must, however, be combined with some kind of multiple contact procedure to improve the efficiency of the separation. Liquid-liquid extraction processes with systematic changes in the properties of the phases are commonly used, but they are as a rule combined with chromatographic processes when high selectivity and sensitivity are desired.

II. REGULATION OF DISTRIBUTION RATIO AND SELECTIVITY

The principles for regulation of the distribution of charged and uncharged organic compounds have been known for several decades, but many of the excellent possibilities for efficient and selective extractions have found a rather limited application outside the analytical field.

A survey of the basic means for distribution control is given in Table 1. Organic compounds are traditionally distributed and separated in uncharged form, and the distribution ratio can then be regulated by the nature of the organic phase, by complexing additives to the liquid phases and by pH, when the compound is an acid or a base. However, for compounds that can be ionized it is often advantageous to make the extraction in that form with the aid of an ion of the opposite charge, a counterion. The additional and unique means for regulation of the distribution are then the kind and the concentration of the counterions, which opens almost unlimited possibilities to variation of the distribution of the charged compounds. Ion-pair extraction has found a wide application in solvent extraction and chromatography (1).

A good knowledge of the parameters that govern the distribution equilibria for ionic compounds is a necessary background to the successful application of the technique. The construction of highly selective or in other respects advanced liquid-liquid extraction processes requires as a rule a more detailed knowledge than chromatographic methods, which can often be developed on the basis of more superficial information.

When a charged compound HA^+ is distributed between an aqueous phase and an organic phase of low polarity with the aid of a counterion X^-, the two ions may form an ion pair in the organic phase, as shown by the formula

$$HA^+_{aq} + X^-_{aq} = HAX_{org} \tag{1}$$

Table 1 Liquid-Liquid Distribution of Organic Compounds

Regulation by	Uncharged (A)	Charged (HA^+)
Organic phase (nature)	+	+
Complexing agent in organic or aqueous phase (kind and concentration)	+	+
pH of aqueous phase (protolytic objects)	+	+
Counterion X^- in aqueous phase (kind and concentration)	0	+

The equilibrium constant of the process $K_{ex}(HAX)$ (the extraction constant), defined by

$$K_{ex}(HAX) = \frac{[HAX]_{org}}{[HA^+]_{aq}[X^-]_{aq}} \tag{2}$$

expresses the combined effect of the phase transfer and the ion-pair formation. If other processes are negligible, the distribution ratio of HA^+, D_{HA}, is given by the relationship

$$D_{HA} = \frac{[HAX]_{org}}{[HA^+]_{aq}} = K_{ex}(HAX)[X^-]_{aq} \tag{3}$$

which shows that the distribution of HA^+ is governed by the extraction constant and the concentration of the counterion in the aqueous phase.

$K_{ex}(HAX)$ changes with the nature of the counterion. An illustration of the width of variation is given in Table 2 (1), which shows that the choice of counterion can change the extraction constant of a cation, tetrabutylammonium, more than 11 logarithmic units. It is obvious that the combination of changes in the nature and concentration of the counterion gives almost unlimited possibilities to variation in D_{HA}. It is of no importance if HA^+ originally is present as a salt with X^- or with another anion, since overall it is only the kind and concentration of anions that are present in the aqueous phase that matters.

Estimation of the conditions for quantitative extraction or separation is easily made if the extraction constants are known. An example is given in Fig. 1, which shows a graphic computation of the separation conditions for two anionic conjugates (2). These are metabolites of a common drug, propranolol, which in the human body is metabolized to a 4'-hydroxy derivative that is conjugated with sulfuric acid or glucuronic acid. The conjugates can be extracted into an organic phase as ion pairs with tetrabutylammonium (TBA). The relationship between the distribution ratio and the counterion concentration [see Eq. (3)] is demonstrated in logarithmic form in the figure. The vertical lines indicate the TBA concentrations that give 99% extraction of the sulfuric acid conjugate and 1% extraction of the glucuronic acid conjugate. It is easily seen that the sulfuric acid conjugate is almost completely extracted at a TBA concentration of 10^{-2} M, whereas the glucuronide is extracted to less than 0.2% under the same conditions. It is obvious that separation by ion-pair extraction is very simple if only those two compounds are present. The glucuronide cannot be quantitatively extracted from the aqueous phase with TBA as counterion, and a more hydrophobic cation with a higher extraction constant is needed.

Ion-pair extraction has also been employed for oxosteroid sulfates in serum (3). After deproteination of the serum sample with methanol and evaporation, the sulfates were extracted with tetrapentylammonium as ion-pairing agent and

Table 2 Extraction Constants of Tetrabutylammonium (Q^+) Ion Pairs (Organic Phase, Chloroform)

Class	Counterion (X^-)	$\log K_{ex(QX)}$
Inorganic	Cl^-	−0.11
	Br^-	1.29
	NO_3^-	1.39
	I^-	3.01
	ClO_4^-	3.48
Carboxylate	Acetate	−2.1
	Phenylacetate	0.27
	3-Hydroxybenzoate	−1.54
	Benzoate	0.39
	Salicylate	2.42
Phenolate	Phenolate	0.05
	Picrate	5.91
	2,4-Dinitro-1-naphtholate	6.45
Sulfonate	Toluene-2-sulfonate	2.33
	Naphthalene-2-sulfonate	3.45
	Anthracene-2-sulfonate	5.11
Sulfate	3-Phenylpropylsulfate	4.20
	2-Naphthylsulfate	4.90
"Acid dye"	Methyl orange	5.47
	Bromothymol blue	8.0
	Dipicrylamine	9.6

benzene as organic solvent. In this case derivatization and reversed-phase liquid chromatography completed the assay. Steroid and bile acid sulfates have been extracted as ion pairs with lucigenin in a flow-injection analysis system (4).

III. EXTRACTION CONSTANT AND ION–PAIR STRUCTURE

The example (Fig. 1) shows that a good estimate of the extraction constant is necessary in all planning of separation systems based on ion-pair extraction. The determination of extraction constants is as a rule fairly simple (5), but evaluation of the results may require some experience since secondary processes can have a significant influence. Tables of extraction constants have been published (6), but detailed conclusions about the influence of structure on the extraction constant are difficult since not only the solvation of the ion-pair components but also the binding between them in the organic phase must be taken into consideration. Present knowledge does not allow the design of prediction systems similar to those used for uncharged molecules, but it is nevertheless possible to draw semiquantitative conclusions of practical importance.

The extraction constant increases with increasing number of alkyl or aryl carbons, and within a homologous series there is usually a linear relationship between the number of methylene groups and the logarithm of the extraction constant. With methylene chloride or chloroform as the organic solvent, log K_{ex} increases by 0.5–0.6 units with the addition of one CH_2. Similar relationships have been found with organic solvents of widely different kinds, which indicates that the change in the constant is mainly due to changes in the interactions in the aqueous phase.

The nature of the ionized groups has a considerable influence on the distribution of the ion pair. The effects depend strongly on the properties of the organic phase. Ions with the same content of alkyl or aryl carbons show the following behavior with the same counterion and CH_2Cl_2 or $CHCl_3$ as the extracting phase.

The extraction constant of organic anions increases in the order carboxylate $<$ sulfonate $<$ sulfate owing to decreasing bonding to the aqueous phase. Hydroxylation in the ortho position to a carboxylate, sulfonate, or sulfate group increases the constant by about 2 log units owing to intramolecular hydrogen bonding.

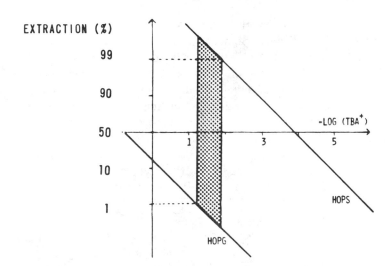

Figure 1 Extraction of propranolol metabolites: HOPS, 4'-hydroxypropranolol sulfate; HOPG, 4'-hydroxypropranolol glucuronide. Organic phase, chloroform. Counterion, tetrabutylammonium (TBA$^+$) Plot: log D = log K_{ex} + log [TBA$^+$] Extraction (%) = $100D/(D + 1)$ (equal phase volumes).

For ion pairs of alkylammonium ions with hydrogen-accepting anions, the extraction constant usually increases in the order quaternary < primary < secondary < tertiary. The hydrogen bonding between cation and anion, which can occur for all alkylammonium ions except the quaternary, makes the ion pair less polar and increases the extraction constant. The effect is strongest for the tertiary ammonium ion, and it decreases with the number of hydrogen substituents at the nitrogen owing to increased bonding to the aqueous phase. The order may be different when the extracting solvent is hydrogen bonding (6).

Hydrophilic substituents, such as hydroxylic, carboxylic, or amino groups, in one of the ion-pair components decrease the extraction constant by 1-2 log units when the organic phase is a halogenated hydrocarbon, but the decrease is much smaller when the extracting solvent is strongly hydrogen bonding, such as pentanol (7). Masking of strongly hydrophilic groups by alkylation or acetylation has a drastic effect on the extraction constant. Methylation of one hydroxy group in morphine increases the extraction constant by 2.4 log units when CH_2Cl_2 is the organic phase, and the acetylation of both hydroxy groups in morphine gives an increase of about 4 log units (8).

IV. INFLUENCE OF ION-PAIR DISSOCIATION AND ASSOCIATION

The extraction of ion pairs can be influenced by different kinds of secondary equilibrium processes (side reactions). Some of these processes, such as protolysis in the aqueous phase and complex formation in the organic phase, can be regulated by the nature and concentration of an external agent and can be used for improving the extent and the selectivity of an extraction procedure. Other secondary equilibria, such as association and dissociation in the organic phase, comprise only the ion pair and can have a positive or negative influence on the extraction process depending on the concentration of the extracted compound. An understanding of these kinds of side reactions is necessary to avoid unexpected disturbances in isolations by the ion-pair extraction technique.

Ion-pair dissociation in the organic phase can have a significant influence on the isolation of a trace component in a sample. The ion pair is an association complex, and its stability depends on the character of the binding forces between the ions and the dielectrical properties of the solvent. The dissociation of the ion pair increases the degree of extraction, as illustrated for a cation HA^+ extracted to an organic phase with X^- as the counterion (9):

$$D_{HA} = K_{ex(HAX)}[X^-]_{aq} + \left(\frac{K_{diss(HAX)} K_{ex(HAX)} [X^-]_{aq}}{[HA^+]_{aq}} \right)^{1/2} \quad . \qquad (4)$$

where the dissociation constant is defined as

$$K_{diss(HAX)} = \frac{[HA^+]_{org}[X^-]_{org}}{[HAX]_{org}} \tag{5}$$

The expression shows that the distribution ratio increases with decreasing concentration of the extracted compound HA^+.

It has been shown that rather drastic extraction improvements can be achieved under favorable conditions (10), and the ion-pair dissociation effect has been used to improve the extraction of hydrophilic organic ions from biological samples. From Eq. (4) follows that a high dissociation constant is advantageous, and in this respect pentanol should be preferable to methylene chloride, as shown in Table 3 (7). Figure 2 shows the extraction of a quaternary ammonium ion as ion pair with dihydrogen phosphate at pH 1-4. The increase in D_{HA} with decreasing concentration of the compound, calculated from the known constants, is indicated by the dotted line. The increase in D_{HA} was, however, much lower and varied with pH owing to the extraction of phosphoric acid to the organic phase (11), which dissociates and suppresses the dissociation of the ion pair. It is obvious that to promote the extraction, it is necessary to use conditions minimizing the coextraction of the counterion to the organic phase.

Association processes, such as dimerization or tetramerization of ion pairs in the organic phase, can also influence the extraction, particularly in the preparative scale. The effect can be illustrated by an expression for the distribution ratio of HA^+ with X^- as counterion when the ion pair HAX is dimerized in the organic phase (9):

$$D_{HA} = K_{ex(HAX)}[X^-]_{aq}(1 + 2K_2(HAX)K_{ex(HAX)}[HA^+]_{aq}[X^-]_{aq}) \tag{6}$$

where the dimerization constant $K_2(HAX)$ is defined as

$$K_2(HAX) = \frac{[H_2 A_2 X_2]_{org}}{[HAX]^2_{org}} \tag{7}$$

From Eq. (6) follows that the distribution ratio increases with increasing concentrations of HA^+. Studies with halogenated hydrocarbons, such as chloroform as

Table 3 Extraction and Dissociation Constants of Choline Picrate (HAX)

Organic phase	Dielectric constant	$\log K_{ex(HAX)}$	$\log K_{diss(HAX)}$
1-Pentanol	13.9	1.19	–3.0
Methylisobutyl ketone	13.1	0.98	–3.3
Ethyl acetate	6.0	0.69	–5.7
Methylene chloride	8.9	–0.12	~–5

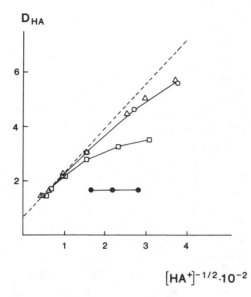

Figure 2 Distribution of quaternary ammonium ion (N-methylamitriptyline = HA^+) as ion pair with dihydrogen phosphate: organic phase, 1-pentanol; aqueous phase, phosphate buffer, $[H_2PO_4^-] = 0.10$. pH 1.0 (●); pH 2.0 (□); pH 3.0 (○); pH 4.0 (△). (From Ref. 11.)

the organic phase, have shown dimerization constants of 10–100 for ion pairs between quaternary ammonium ions and halides or sulfonates, whereas ion pairs of primary and secondary ammonium ions give $K_2(HAX) = 10^3$–10^4. The higher dimerization tendency of the latter compounds may be due to their hydrogen-donating ability (6). Overall, it is obvious that the tendency to ion-pair association increases with decreasing dielectrical constant of the organic phase.

V. MECHANISM OF ION–PAIR EXTRACTION

The ion-pair extraction process includes at least two steps: the transfer between the phases and the formation of the ion pair. Mechanistic features of the process, such as the site for the formation of the ion pairs, cannot be elucidated by equilibrium expressions, and measurements of the transfer rates of the ion-pair components between the phases are needed to obtain a better understanding.

Studies of this kind have so far been performed with a limited number of systems (12). The observations indicate that the extraction process comprises a transfer of the compounds in ionic form to the organic phase, where the tendency to ion-pair formation is higher owing to the lower polarity of the medium. The measurements showed drastic differences in ion-pair formation constants K_a in the two phases, K_a usually being of the order 20–50 in water but 20,000 or more in methylene chloride (6).

Generalizations of these findings to all kinds of systems may be dangerous. A high degree of ion-pair formation in the aqueous phase was observed between highly hydrophobic components with structural features that promote a strong hydrophobic interaction (13). Other kinds of extraction mechanisms than those already mentioned may then play an important role.

A further comment is necessary. Studies on the distribution of charged organic compounds have mainly been concentrated on the determination of equilibrium constants and related data, and mechanistic discussions have usually been avoided. The technique was designated "ion-pair extraction" since the early results indicated a high degree of ion-pair formation in the organic phase under regular extraction conditions. "Counterion-aided extraction" might be a more correct although too complicated designation. However, it must be emphasized that ion-pair formation in the aqueous phase is by no means an indispensable first step in the extraction process. Instead, the fact is that a high degree of ion-pair formation in the aqueous phase decreases the distribution ratio and limits the possibilities of influencing the extraction by the counterion concentration (13).

VI. INFLUENCE OF PROTOLYSIS

Acids and bases acquire on protolysis a change in charge that strongly affects their distribution in uncharged form and as ion pairs. Change in the pH of the aqueous phase is a common way to regulate the distribution of protolytic compounds. The technique has furthermore the advantage that an exact calculation of the distribution can be made since the relevant equilibrium constants are often easy to determine.

The extraction of a weak base as ion pair and in uncharged form can be used as an illustration. The weak base A is extracted from an aqueous phase containing the aprotic anion X^-. At low pH the base is present as HA^+, and it is extracted entirely as the ion pair HAX. At higher pH, HA^+ is protolyzed to A and can then be extracted to the organic phase in that form. When a simultaneous extraction of the ion pair HAX and the base A occurs, the distribution ratio for A has the following form (9):

$$D_A = \frac{[A]_{org} + [HAX]_{org}}{[A]_{aq} + [HA^+]_{aq}} = \frac{K_{D(A)}K'_{HA}}{K'_{HA} + a_H^+} + \frac{K_{ex(HAX)}[X^-]_{aq}a_H^+}{K'_{HA} + a_H^+} \qquad (8)$$

where $K_D(A)$ is the distribution constant of A and $K'HA$ is the acid dissociation constant of HA^+. A graphic illustration of Eq. (8) is given in Fig. 3. With the

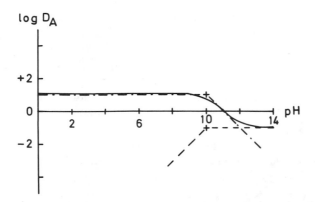

Figure 3 Distribution of amine extractable as ion pair and base. Counterion concentration $[X^-] = 0.1$; $\log K_{ex\,(HAX)} = 2.0$; $\log K_{D(A)} = -1$; $pK'_{HA} = 10.0$.

constants and conditions used for the construction, ion-pair extraction dominates at a pH below 11 and extraction as base at a pH above 13.

VII.   COMPLEXATION IN THE ORGANIC PHASE

Ion-pair extraction can also be combined with complexation in the organic phase. If a cation HA^+ is extracted with a counterion X^- in the aqueous phase and a complexing agent S in the organic, ion-pair adducts are formed in the organic phase as shown by the formula

$$HA^+_{aq} + X^-_{aq} + nS_{org} = HAXS_{n,org} \tag{9}$$

If forms of adducts other than $HAXS_n$ are negligible, the distribution ratio of HA^+ (9) is given by

$$D_{HA} = K_{ex(HAX)}K_n[X^-]_{aq}[S]^n_{org} \tag{10}$$

where

$$K_n = \frac{[HAXS_n]_{org}}{[HAX]_{org}[S]^n_{org}} \tag{11}$$

D_{HA} can be controlled by the nature and the concentration of X^- and S. The influence of complexation on ion-pair extraction has been studied in a number of systems (6), mainly with lipophilic alcohols as the complexing agent, and a constant value of n was found in most cases.

The strongly hydrogen accepting trioctylphosphine oxide (TOPO) has an interesting influence on the extraction of ion pairs of organic ammonium ions

with different degrees of substitution, as shown in Fig. 4 (14). Increase in the
TOPO concentration in the organic phase enhances the extraction of ion pairs
of primary and secondary ammonium ions, whereas the extraction of tertiary
and quaternary ammonium ion pairs is almost unaffected. This indicates that
only the first two kinds of ion pairs have any hydrogens available for binding to
TOPO.

A considerable change in the extraction selectivity can be obtained by the use
of structure-specific adduct-forming agents, such as cyclic polyethers. Dibenzo-
18-crown-6 gives highly stable 1:1 complexes with ion pairs containing primary
ammonium ions, which fit well in the ring structure of the complexing agent
and are bound by hydrogen bonds to the oxygens. Such a good fit cannot be
obtained when the ammonium ion has a higher degree of substitution and the
complex constant is therefore much smaller. The effect of complexation by the
crown ether on two cations, octylammonium and trimethylnonylammonium, is
demonstrated in Fig. 5 (15). The cations are extracted into chloroform as ion
pairs with salicylate. Addition of the crown ether to the organic phase gives a
strong increase in the distribution ratio of the primary ammonium ion; the
quaternary ammonium is affected to a much smaller extent.

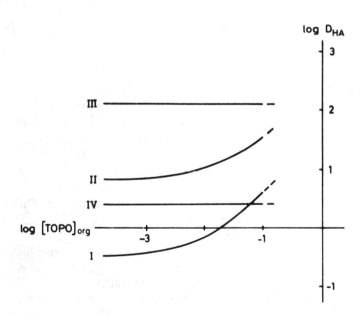

Figure 4 Trioctylphosphine oxide (TOPO) as complexing agent: organic phase,
TOPO in chloroform; counterion, naphthalene-2-sulfonate, 0.05 M. Substrates:
I, octylammonium; II, N-methyloctylammonium; III, N,N-dimethyloctylam-
monium; IV, N,N,N-trimethyloctylammonium.

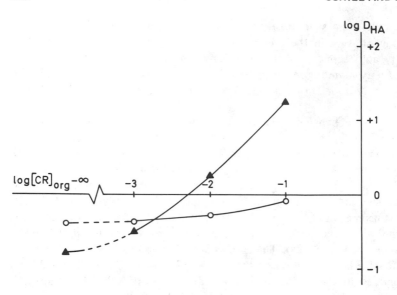

Figure 5 Dibenzo-18-crown-6 (CR) as complexing agent: organic phase, CR in chloroform; counterion, naphthalene-2-sulfonate, 0.05 M. Substrates: (▲) octylammonium; (○) N,N,N-trimethyloctylammonium.

VIII. EXTRACTION OF HYDROPHILIC IONS

The basic approach to the extraction of hydrophilic ions is very simple: use a counterion of such hydrophobicity that an extraction of sufficient magnitude is obtained and in such a concentration that the distribution ratio (i.e., the product of the extraction constant and the counterion concentration) is sufficient for a quantitative extraction. However, to attain a distribution ratio of 100, which is the requirement for quantitative extraction with equal phase volumes, can often be difficult. The maximum concentration of the counterion decreases when its hydrophobicity increases owing to a limited solubility in the aqueous phase. The highly hydrophobic counterions are furthermore prone to extraction as ion pairs with buffer components of other hydrophilic ions in the same aqueous solution as the compound, making the procedure unselective. Extractions from complex samples, such as biological material, must as a rule be followed by further isolation procedures of chromatographic or other nature as exemplified here.

Urinary metabolites of caffeine and theophylline have been extracted at pH 11 as ion pairs with tetrabutylammonium and a mixture of ethylacetate, chloroform, and isopropanol as organic phase (16). This extraction procedure combined with liquid chromatographic determination was compared with direct chromatography of diluted urine samples and was recommended when a more precise estimation was required.

Tianeptine, an antidepressant drug with a carboxylic moiety, was extracted from biological samples at pH 7 as an ion pair with tetraheptylammonium and heptane with 2% octanol as organic solvent (17). After back-extraction to an aqueous phase an aliquot of this was injected into the liquid chromatograph. Tetraphenylborate has been used as ion pairing anion and diethyl ether as solvent in the measurement of tyramine in human plasma by liquid chromatography and amperometric detection (18).

Acetylcholine (a hydrophilic quaternary ammonium ion) can be extracted with 3,5-di-*t*-butyl-2-hydroxybenzene sulfonate as counterion and methylene chloride as the extracting solvent (19). The extraction is made under such conditions that the calculated distribution ratio is just about unity. However, acetylcholine is present in the biological material in such a low concentration that ion-pair dissociation in the organic phase increases the distribution ratio to such an extent that a quantitative yield is obtained after two extractions of the biological material. The same sulfonate counterion has been used in a method for apomorphine in rat brain to enable ion-pair extraction to a small volume of methylene chloride prior to liquid chromatography (20).

Methylguanidine is so hydrophilic that even the extremely hydrophobic counterion dipicrylamine (see Table 2) gives too low a distribution ratio for quantitative recovery with batch extraction to methylene chloride. However, column extraction with kieselguhr as support for the aqueous phase gives quantitative extraction with a fairly limited volume of organic phase (21).

A more effective way of increasing the extraction of a hydrophilic ion is to use proteolytic, hydrophobic agents that can act as counterions and simultaneously as complexing agents in uncharged form. The principle can be illustrated by the following formula, where HA^+ is the object and HX the extracting agent:

$$HA^+_{aq} + nHX_{org} = HAX(HX)_{n-1,org} + H^+ \tag{12}$$

The expression for the distribution ratio of HA^+ is then

$$D_{HA} = \frac{K_{ex(HAX)}K_{n-1}[HX]^n_{org}K'_{HX}}{K_{D(HX)}a_{H^+}} \tag{13}$$

K'_{HX} is the acid dissociation constant and $K_{D(HX)}$ the distribution constant of HX. K_{n-1} is the formation constant of the ion-pair complex between HAX and HX [see Eq. (11)]. The expression shows that the extraction can be optimized by a proper choice of extracting agent concentration and pH. One of the most popular extracting agents of this kind is bis-(2-ethylhexyl)phosphoric acid (HDEHP), which is highly hydrophobic and suitable for extraction of hydrophilic ions as catecholamines (22). The extraction of structurally related sympathomimetics, such as ephedrine, isoprenaline, and terbutaline, has been studied employing HDEHP and other anionic counterions and different organic phases

(23). This kind of extraction procedure has also been applied to biological material, but it is unselective and must be combined with further isolation procedures of a chromatographic character (24).

An improvement in the selectivity of the extraction procedure can sometimes be achieved by using a selective complexing agent in combination with a counterion of suitable hydrophobicity. An example of a process of this kind is the use of diphenylborate as complexing agent by extraction of catecholamines (25). The complexation gives an anionic diol-borate complex,

$$\tag{14}$$

which is extracted to the organic phase with a hydrophobic quaternary ammonium ion.

Ion-pair extraction with tetraoctylammonium bromide in the presence of diphenylborate has also been used for liquid chromatographic determination of biogenic amines in brain tissue samples (26). The organic solvent was heptane containing 1% octanol. Dihydroxyphenylalanine in blood plasma was extracted into hexanol after complexation with diphenylborate and ion-pair formation with tetrapentylammonium (27).

IX. ION–PAIR CHROMATOGRAPHY

The demand for separations and isolations of structurally closely related compounds cannot be fulfilled by liquid-liquid extraction processes alone, despite the use of highly selective reagents and repeated extractions with a systematic change in the properties of the phases. Chromatographic procedures are often needed to improve the separating efficiency, and such methods have dominated on the submicromolar level during the last 15 years. The application of the chromatographic procedures to preparative-scale separations has also increased lately with the development of new technical facilities.

The technique to distribute charged compounds with the aid of counterions is applicable to liquid-liquid and liquid-solid chromatography, and it has a wide application in separation science under the designation of ion-pair chromatography. The distribution of the charged analyte between the mobile and the stationary phases in the chromatographic system is in principle governed in the same way as in liquid-liquid extraction, but different theoretical approaches must be used when the stationary phase is a liquid or an adsorbing solid surface.

A. Liquid-Liquid Systems

The separation of charged compounds in genuine liquid-liquid systems is based on the principles used in solvent extraction systems as summarized in the following expressions for the capacity ratio k' of a cationic compound HA^+, with X^- as counterion in the aqueous phase (28). Stationary aqueous phase (*straight-phase systems*):

$$k'_{HA} = \frac{1}{K_{ex(HAX)}[X^-]_{aq}} \frac{V_s}{V_m} \tag{15}$$

Stationary organic phase (*reversed-phase systems*):

$$k'_{HA} = K_{ex(HAX)}[X^-]_{aq} \frac{V_s}{V_m} \tag{16}$$

The capacity ratio is defined as

$$k' = \frac{V_r - V_m}{V_m} \tag{17}$$

where V_r and V_m are the volumes of mobile phase needed to elute a retained and a nonretained compound, respectively. V_s/V_m is the volume ratio of the stationary and mobile phases in the chromatographic column. In the straight-phase system, the compound migrates as an ion pair; in the reversed phase system, it is retained as an ion pair.

Systems with a stationary aqueous phase and a mobile organic phase have been widely used in biomedical analysis. They are suitable for fairly hydrophobic compounds that can be extracted into an organic phase before introduction into the chromatographic system. An example of the separation selectivity that can be obtained is given in Fig. 6, which shows a separation of possible metabolites of a drug substance, alprenolol, with perchlorate as counterion (29). A complete separation of the five hydroxy derivatives (indicated by arrows in the figure) is easily obtained. Careful thermostating of the whole chromatographic system is a prerequisite for the long-term stability of systems of this kind.

In reversed-phase systems, the organic stationary phase is coated on a hydrophobic adsorbent. The necessity of long-term stability of the chromatographic system has made the choice of the organic liquid phase fairly limited. Interesting results have been obtained with the strongly hydrogen-accepting tributyl-phosphate (TBP) as the stationary liquid. Highly hydrophilic ions are retained well in systems of this kind, as demonstrated in Fig. 7, which shows a separation of catecholamines and related amino acids as ion pairs with perchlorate (30). The system is highly stable owing to the low solubility of TBP in the aqueous mobile phase.

B. Liquid-Solid Systems

The dominating part of all ion-pair chromatographic separations is performed in the liquid-solid mode with a hydrophobic adsorbent and an aqueous mobile

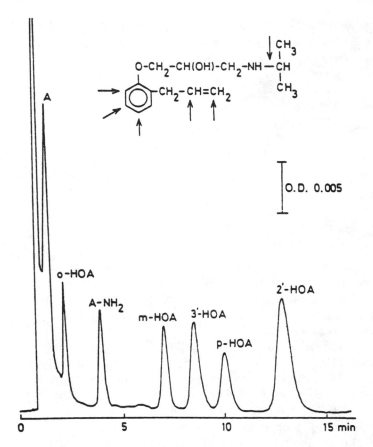

Figure 6 Separation of possible metabolites of alprenolol. Stationary phase, 0.8 M NaClO₄ + 0.2 M HClO₄ on Partisil 10 (porous silica, 10 μm). Mobile phase, 1-butanol + 1,2-dichloroethane + hexane (8:32:60). Substrates: A, alprenolol; A–NH₂, desisopropylalprenolol; HOA, hydroxy-substituted alprenolol (arrow indicates OH position).

phase. These systems have the advantage of being compatible with aqueous samples and the flexibility is, owing to the ion-pair distribution principle, almost unlimited. They can be adapted to widely different kinds of compounds, hydrophilic as well as hydrophobic.

The main difference between the liquid-solid and the liquid-liquid systems is that, contrary to a liquid stationary phase, a solid adsorbent has a limited binding capacity. This opens new possibilities to regulation of retention and selectivity. The principles for retention can be summarized as follows:

1. The binding of the substrate HA^+ is accompanied by displacement of a system ion of the same charge Q^+ or binding of a counterion X^-.
2. The substrate and the mobile-phase ions compete for the binding capacity of the adsorbent.

The retention, as a rule, follows models based on the assumption of a Langmuir adsorption (31,32), as demonstrated by

$$k'_{HA} = \frac{K^0 q K_{HAX}[X^-]_{aq}}{1 + K_{QX}[Q^+]_{aq}[X^-]_{aq}} \quad (18)$$

where K^0 is the capacity of the adsorbent and q is the phase ratio in the chromatographic system. K_{HAX} and K_{QX} are defined in analogy with Eq. (2).

The expression shows that the retention can be regulated by the concentration and the nature of the counterion as in all ion-pair distribution techniques. The increase in the counterion concentration, however, also increases the denominator, which results in a curved relationship between k' and the counterion concentration. A mobile-phase ion (Q^+) with the same charge as the substrate also influences the retention, and an increase in its hydrophobicity and concentration decreases the retention of the substrate. The competing effect of this ion is of great practical importance for the proper functioning of the

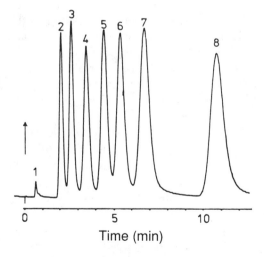

Time (min)

Figure 7 Separation of catecholamines as ion pairs. Stationary phase, tributyl-phosphate (TBP) on LiChrosorb RP-8 (alkyl-bonded, porous silica, 10 μm). Mobile phase, 0.1 M NaClO$_4$ + 0.3% TBP, pH 2.1. Substrates: (2) epinephrine; (3) norepinephrine; (4) dopa; (5) a-methylnorepinephrine; (6) dopamine; (7) a-methyldopa; (8) a-methyldopamine.

chromatographic system since an increase in the competition is an effective way to eliminate overloading effects in the system (32).

A typical example of the application of reversed-phase ion-pair chromatography to the isolation of organic compounds from biological material is shown in Fig. 8. The compounds of primary interest in the rat brain sample were two amines, dopamine and epinine, and two carboxylic acid derivatives, HVA and DOPAC (33). The compounds are retained by a hydrophobic adsorbent in acid medium, the amines as ion pairs and the acids in uncharged form. The separation is incomplete in buffer solution, but addition of a hydrophobic anion, hexylsulfate, increases the retention of the amines while the acids are almost unaffected, and complete separation can be achieved. The separation is very good despite the simple pretreatment of the biological material, homogenization in aqueous perchloric acid.

X. SEPARATION OF ENANTIOMERIC IONS

Optical isomers (enantiomers) of biologically active compounds often have widely different pharmacological effect, and isolation of single enantiomers from biological material is of great interest in many biomedical studies. The rapid development of separation science during the last years has given a series of new methods for enantioselective separations.

Enantiomeric compounds cannot be separated in unchanged form in normal distribution systems, but a separation can be achieved if they are combined with another enantiomeric compound (a chiral selector) of such structure that the two diastereomers formed have different distribution properties. The chiral selector can be coupled to the enantiomeric objects by covalent bonding, but separation systems based on the formation of diastereomeric complexes with reversible bonding often have distinct advantages. Several of these systems for enantiomer separation are based on ion-pair distribution.

Most of the systems give fairly low differences in distribution ratio between the diastereomeric complexes, and a chromatographic technique is without exception necessary. Ion-pair distribution systems containing an uncharged chiral selector have been used in several cases. Cram et al. (34) have made separations of primary amines with chiral crown compounds as selectors, and tartaric acid esters have been used as selectors for enantiomeric amino alcohols by Prelog et al. (35) and others.

Enantiomeric ions can also be separated with an enantiomeric counterion as the chiral selector. Three interaction points between the enantiomeric substrate and the selector seem to be needed, and as a rule the chiral counterions have a rigid or bulky structure with a charged group and a hydrogen-bonding group in the vicinity of the asymmetric center. The separations are usually made in chromatographic systems, with a hydrophilic adsorbent and the chiral

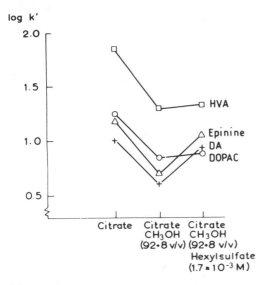

(a)

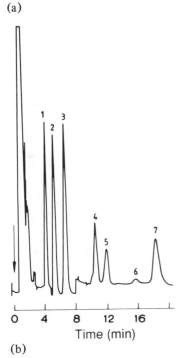

(b)

Figure 8 (a) Optimization of capacity ratios. Stationary phase, Nucleosil C₁₈ (alkyl-bonded, porous silica, 5 μm). Mobile phase, citrate buffer, pH 4.2, + additives. Substrates, HVA, homovanillic acid; epinine, N-methyldopamine; DA, dopamine; DOPAC, 3,4-dihydroxyphenylacetic acid. (b) Chromatogram from a rat striatum sample. Conditions as in a. Substrates: (1) DOPAC; (2) DA; (3) epinine; (4) 5-HIAA; (5) HVA; (6) 3-methoxytyramine; (7) serotonin.

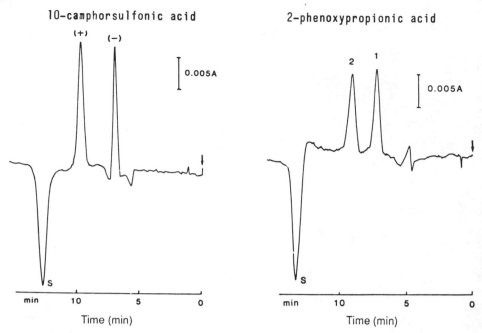

Figure 9 Separation of enantiomeric acids. Stationary phase, LiChrosorb DIOL (diol-bonded, porous silica). Mobile phase, quinine acetate, 3.5×10^{-4} M in methylene chloride + 1-pentanol (99:1). (From Ref. 36.)

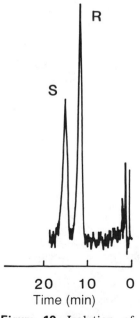

Figure 10 Isolation of R- and S-propranolol in plasma. Stationary phase, LiChrosorb DIOL. Mobile phase, 2.5 mM ZGP + 0.2 mM triethylamine + 500 ppm H_2O in methylene chloride. (From Ref. 37.)

counterion added to an organic mobile phase that should have low polarity to improve the ion-pair formation. Pettersson et al. (36,37) have developed a technique of this kind for the separation of cationic and anionic enantiomers. Quinine and other amine alcohols have been used for the separation of hydrogen-bonding carboxylic acids, as illustrated in Fig. 9. A dipeptide, N-benzoxycarbonylglycyl-L-proline (ZGP), is suitable as counterion for separation of enantiomeric amino alcohols, as demonstrated for a drug, propranolol, in a plasma sample in Fig. 10.

REFERENCES

1. Schill, G., Ehrsson, H., Vessman, J., and Westerlund, D., *Separation Methods for Drugs and Related Organic Compounds*, 2nd ed. Swedish Pharmaceutical Press, Stockholm (1984).
2. Wingstrand, K. H., and Walle, T., Isolation from urine of 4'-hydroxypropranolol sulfate, a major new propranolol metabolite, by ion-pair chromatography, *J. Chromatogr.* 305:250–255 (1984).
3. Iwata, J., and Suga, T., Determination of 17-oxosteroid sulphates in serum by ion-pair extraction, prelabelling with dansylhydrazine and high-performance liquid chromatography with fluorescence detection, *J. Chromatogr.* 474:363–371 (1989).
4. Maeda, M., and Tsuji, A., Fluorescence and chemiluminescence determination of steroid and bile acid sulphates with lucigenin by flow injection analysis based on ion-pair solvent extraction, *Analyst* 110:665–668 (1985).
5. Modin, R., and Schill, G., Quantitative determinations by ion pair extraction. Part 6. Principles for determination of extraction constants. *Acta Pharm. Suec.* 7:585–614 (1970).
6. Schill, G., Isolation of drugs and related organic compounds by ion-pair extraction, in: *Ion Exchange and Solvent Extraction* (J. A. Marinsky and Y. Marcus, eds.), Vol. 6. Marcel Dekker, New York, pp. 1–54 (1974).
7. Modin, R., and Bäck, S., Quantitative determinations by ion pair extraction. Part 10. Extraction of picrate ion pairs into polar organic solvents. *Acta Pharm. Suec.* 8:585–590 (1971).
8. Schill, G., Photometric determination of amines and quaternary ammonium compounds with bromothymol blue. Part 4. Extraction constants and calculation of extraction conditions. *Acta Pharm. Suec.* 2:13–46 (1965).
9. Schill, G., Ehrsson, H., Vessman, J., and Westerlund, D., *Separation Methods for Drugs and Related Organic Compounds*, 2nd ed. Swedish Pharmaceutical Press, Stockholm, Chapter 1, pp. 1–31 (1984).
10. Lagerström, P. O., Borg, K. O., and Westerlund, D., Fluorimetric determinations by ion pair extraction. Part 4. Studies on quantitative ion pair extraction of ammonium compounds with anthracene-2-sulphonate as counter ion. *Acta Pharm. Suec.* 9:53–62 (1972).
11. Kolstad, A. K., Distribution of organic ammonium ions to 1-pentanol as ion pairs with dihydrogen phosphate. I. Effect of buffer composition on distribution ratios, *Acta Pharm. Suec.* 20:1–10 (1983).

12. Nordgren, T., and Sjödén, E. K., Kinetic studies on ion-pair extraction. II. Ion-pair extraction of *N,N*-dimethylprotriptyline. The effect of ion-pair formation in the organic phase. *Acta Pharm. Suec.* 15:241–254 (1978).

13. Fransson, B., and Schill, G., Isolation of acidic conjugates by ion-pair extraction. I. Extraction of glycine, glucuronic and sulphuric acid conjugates, *Acta Pharm. Suec.* 12:107–118 (1975).

14. Schröder-Nielsen, M., Quantitative determinations by ion-pair extraction. Part 11. Extraction of carboxylates and sulphonates as ion pairs and adducts with dibenzo-18-crown-6 and other adduct-forming agents, *Acta Pharm. Suec.* 11:541–562 (1974).

15. Schröder-Nielsen, M., Quantitative determinations by ion-pair extraction. Part 13. Extraction of salicylate and 2-naphthalenesulfonate as ion pairs and adducts with trioctylphosphine oxide, *Acta Pharm. Suec.* 13:145–156 (1976).

16. Scott, N. R., Chakraborty, J., and Marks, V., Determination of the urinary metabolites of caffeine and theophylline by high-performance liquid chromatography. A comparative study of a direct injection and an ion-pair extraction procedure, *J. Chromatogr.* 375:321–329 (1986).

17. Nicot, G., Lachatre, G., Gonnet, C., Mallon, J., and Mocaer, E., Ion-pair extraction and high-performance liquid chromatographic determination of tianeptine and its metabolites in human plasma, urine and tissues, *J. Chromatogr.* 381:115–126 (1986).

18. Causon, R. C., and Brown, M. J., Measurement of tyramine in human plasma, utilising ion-pair extraction and high-performance liquid chromatography with amperometric detection, *J. Chromatogr.* 310:11–17 (1984).

19. Ulin, B., Gustavii, K., and Persson, B.-A., Bioanalysis of picomole amounts of acetylcholine by ion-pair partition chromatography applied to rat sciatic nerve, *J. Pharm. Pharmacol.* 28:672–675 (1976).

20. Eriksson, B-M., Persson, B-A. and Lindberg, M., Determination of apomorphine in plasma and brain tissue by ion-pair extraction and liquid chromatography, *J. Chromatogr.* 185:575–581 (1979).

21. Eksborg, S., Persson, B-A., Allgén, L-G., Bergström, J., Zimmerman, L., and Fürst, P., A selective method for determination of methylguanidine in biological fluids. Its application in normal subjects and uremic patients, *Clin. Chim. Acta* 82:141–150 (1978).

22. Modin, R., and Johansson, M., Quantitative determinations by ion pair extraction. Part 8. Extraction of aminophenols and aminoalcohols by bis-(2-ethylhexyl)phosphoric acid. *Acta Pharm. Suec.* 8:561–572 (1971).

23. Brandts, P. M., Maes, R. A. A., Leferink, J. G., de Ligny, C. L., and Nieuwdorp, G. H. E., Ion-pair extraction of some sympathomimetics. Description of an extraction model for terbutaline and investigation of some factors influencing the recovery of sympathomimetics, *Anal. Chim. Acta* 135: 85–98 (1982).

24. Eriksson, B-M., Andersson, I., Borg, K. O., and Persson, B-A., Determination of adrenaline and noradrenaline in plasma by an isotope derivative method and ion pair liquid chromatography, *Acta Pharm. Suec.* 14:451–458 (1977).

25. Smedes, F., Kraak, J-C., and Poppe, H., Simple and fast solvent extraction system for selective and quantitative isolation of adrenaline, noradrenaline and dopamine from plasma and urine, *J. Chromatogr.* 231:25–39 (1982).

26. Herregodts, P., Michotte, Y., and Ebinger, G., Combined ion-pair extraction and high-performance liquid chromatography for the determination of the biogenic amines and their major metabolites in single brain tissue samples, *J. Chromatogr.* 421:51–60 (1987).

27. Tsuchiya, H., and Hayashi, T., Determination of L-3,4-dihydroxyphenylalanine in blood by high-performance liquid chromatography after solvent extraction, *J. Chromatogr.* 491:291–298 (1989).

28. Eksborg, S., and Schill, G., Ion pair partition chromatography of organic ammonium compounds, *Anal. Chem.* 45:2092–2100 (1973).

29. Borg, K. O., unpublished communication

30. Jansson, H. J. L., Tjaden, U. R., de Jong, H. J., and Wahlund, K-G., Reversed-phase ion-pair chromatography of biogenic catecholamines and their a-methyl homologues with tributylphosphate as stationary phase, *J. Chromatogr.* 202:223–232 (1980).

31. Tilly Melin, A., Askemark, Y., Wahlund, K.-G., and Schill, G., Retention behaviour of carboxylic acids and their quaternary ammonium ion pairs in reversed phase chromatography with acetonitrile as organic modifier in the mobile phase, *Anal. Chem.* 51:976–983 (1979).

32. Sokolowski, A., and Wahlund, K-G., Peak tailing and retention behaviour of tricyclic antidepressant amines and related hydrophobic ammonium compounds in reversed phase ion-pair liquid chromatography on alkyl-bonded phases, *J. Chromatogr.* 189:299–316 (1980).

33. Magnusson, O., Nilsson, L. B., and Westerlund, D., Simultaneous determination of dopamine, DOPAC and homovanillic acid. Direct injection of supernatants from brain tissue homogenates in a liquid chromatography-electrochemical detection system. *J. Chromatogr.* 221:237–247 (1980).

34. Chao, Y., Weisman, G., Sogah, G., and Cram, D., Host-guest complexation. 21. Catalysis and chiral recognition through designed complexation of transition states in transacylations of amino ester salts, *J. Amer. Chem. Soc.* 101:4948–4958 (1979).

35. Prelog, V., Stojanac, Z., and Kovacevic, K., Separation of enantiomers by partition between liquid phases. *Helv. Chim. Acta* 65:377–384 (1982).

36. Pettersson, C., and No, K., Chiral resolution of carboxylic and sulphonic acids by ion-pair chromatography, *J. Chromatogr.* 282:671–684 (1983).

37. Pettersson, C., and Josefsson, M., Chiral separation of aminoalcohols by ion-pair chromatography. *Chromatographia* 21:321–326 (1986).

13

The Biostil Process

Lars Ehnström, Jonas Frisenfelt, and Magnus Danielsson

Chematur AB
Karlskoga, Sweden

I. INTRODUCTION

The fermentation process BIOSTIL was developed at the end of the 1970s for continuous production of ethanol. BIOSTIL can be said to be an extractive bioconversion process in which a continuous extraction of the product and other inhibiting substances is made with a centrifugal separator in combination with a distillation procedure. The process is characterized by its ability to anaerobically ferment highly concentrated substrates in a single-tank fermentor with a continuous extraction of the product and other inhibiting substances. Each unit operation includes robust and well-tested equipment and techniques, ensuring that the industry's high demands on reliability of operation and accessibility are met. Since the introduction in 1982, 20 plants have been put into operation.

II. FERMENTATION INDUSTRY

Fermentation has great chances of becoming a future production technique, provided that industry's stringent requirements with regard to energy consumption, environmental impact, and accessibility can be met in a cost-effective manner. Yield and energy, based on theory, can be checked and optimized in laboratory and pilot operations. Accessibility, which in most cases has the greatest influence on the economy, is established by industrial experience extending over a long period. A great advantage of fermentation is that it is carried out at a low temperature, which creates opportunities for good energy economy; since water is usually the solvent, it is possible to meet strict environmental restrictions.

Unfortunately, the advantages cannot be utilized because it is necessary to work with very dilute solutions in fermentation systems. The concentration of the product must be kept low because it has an inhibiting effect on the production rate. It is, however, desirable that the end product be concentrated and therefore the added dilution water must be removed by means of an energy-consuming method. This holds true for old established industry producing alcohol and yeast, as well as for the modern pharmaceutical industry and the latest cultivation systems with recombinant (rDNA) cells.

Almost all new plants in production today are based on batch technology, despite that different continuous systems have been available. Single-tank, cascade or tower plants, have not convinced the industry to change direction. Among other things, the infection risks have had a deterring effect. The only technique being used to a large extent is Melle Boanot, which was suggested as early as 1936. This technique reduced the fermentation time for ethanol from 30-40 h to about 15 h. (Two excellent reviews of ethanol fermentation technologies are referred to in Refs. 1 and 2.)

The industry needs more than increased fermentor productivity to justify a continuous fermentation. What is more important for the economy is, as already mentioned,

Accessibility
Energy consumption
Waste (environmental) handling
Yield

The value of a continuous process depends on how well these criteria can be fulfilled.

A. Accessibility

For production of ethanol, for example, a robust, reliable process is required. Proven techniques and equipment should be used in each unit operation. The economic improvements obtained, such as higher yield or a lower energy consumption, disappear unless the accessibility of the plant is high. Infection can be devastating to the process, and the plant must therefore be resistant to infection.

B. Environmental

Effluent problems are in most cases a plague to the distillery and fermentation industry. As a rule conventional distilleries produce about 10 times as much slop as product. As slop has extremely high BOD (biological oxygen demands) and poor color characteristics, the costs for minimizing the environmental impact are very high. New plants or extensions are not granted a production permit unless satisfactory solutions to the waste problems can be demonstrated.

C. Yield

The yield of a fermentation process can be improved in the following ways:

By working under continuous conditions, losses are reduced with regular cleaning.

By reducing the by-product production, for example glycerol in ethanol fermentation.

By decreasing the physical sugar (substrate) losses in the waste streams by reducing the waste volume.

By using less substrate for production of cells, that is, by reusing cells by recirculation.

To fulfill these demands, a process was developed for ethanol fermentation in which the fermentor was kept under a reduced pressure so that alcohol could be stripped off at fermentation temperatures. The technique functioned well in the laboratory but has not been commercialized. The main problem is the large volumes of CO_2 formed during fermentation, which must be compressed to atmospheric pressure. These and several other problems already mentioned have been solved in a satisfactory way in the BIOSTIL process (3-7).

III. BIOSTIL PROCESS

A. Basic Concept

The method has been developed for anaerobic alcohol fermentation but can also be used for the production of other biochemical substances or cleaning processes. In a process loop consisting of a single fermentor and an extraction section according to Fig. 1, a steady state is established. The steady state in the fermentor is maintained by a continuous addition of substrate with a high concentration of fermentable material and continuous removal of fermentation liquid containing inhibiting substances at a concentration below the limit for operating disturbances.

The liquid from the fermentor is separated in a centrifugal separator into a cell concentrate phase and a mainly cell-free phase. The concentrate phase is brought back to the fermentor to secure high cell density and a high fermentation rate without substrate or yeast losses.

The cell-free phase is separated in a product-concentrated flow leaving the process loop and a residual flow. The main part of this is pasteurized and

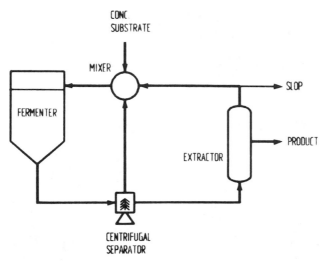

Figure 1 BIOSTIL: basic concept.

brought back to the fermentor; the rest is used for regulating the material balance by drainage to the waste section.

When the product consists of alcohol, distillation or stripping is the best method of separation, and then the desired pasteurization takes place simultaneously (see later). Alternatively, liquid-liquid extraction, a membrane technique, or various sorption methods can be used.

Cell separation is best carried out in a high-speed disk-stack centrifuge (separator), as this machine can be regulated so that a desirable classification of particles is obtained. This is very important to maintain a stable equilibrium for solids, mainly proteins and fibers, in the system, and to separate and prevent the concentration of bacteria in the fermentor, (see later).

Since the fermentation product is removed continuously at the same rate as it is formed, its inhibiting effect on the microorganism is avoided, as already mentioned. The fermentation is therefore no longer limited by the product concentration but instead by other factors. In substrates fermentation with a material like molasses or other sugar concentrates, the total osmotic pressure limits the production ability of the cells. The feedstock concentration can therefore be increased until the osmotic pressure in the fermentor has reached a limit for securing net yeast growth. In this respect the process is unique among commercial fermentation plants. This is illustrated in Fig. 2, where the maximum substrate concentration is plotted against F/N, a factor describing the composition of the substrate (for calculation of F/N, see Table 2). This factor indicates the ratio between fermentable and nonfermentable materials. With substrates containing higher concentrations of fermentable material ($F/N > 3.5$), the viscosity and pumpability of the slop or effluent are the limiting factors. This is also the case when the substrate has a high content of suspended material, which is common for so-called whole mash from grain.

B. Centrifugal Separation

Separation of particles and active microorganisms, as well as proteins and fibers, is a key operation in the process. This is especially valid for equipment that is directly connected to the fermentor, which must fulfill the following demands:

Robust and reliable in operation
Small holdup
Sanitary and easy to clean
Must be able to classify (fractionate) particles

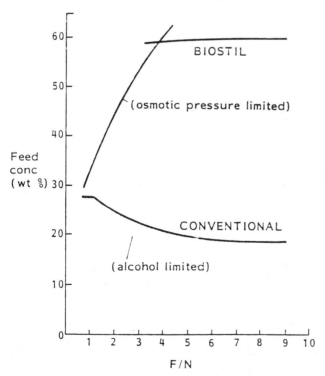

Figure 2 Maximum fermentable sugar concentration of the substrate.

The last point is especially necessary for satisfactory functioning of the process. The machine that best fulfills these demands is a high-speed centrifugal separator with nozzles in the periphery according to Fig. 3.

A centrifugal separator (8) can be seen as a sedimentation basin in which the sedimentation speed has been increased by centrifigual force. The relative centrifugal force for large industrial machines lies in the range of 10,000 $\times$ g. The sedimentation surface has been optimized by installing conical disks, the so-called disk stack, and area equivalents of around 100,000 m^2 are common. (The area equivalent is the surface area required in a sedimentation tank to achieve the same result as in a centrifugal separator.) For separation of microorganisms, so-called nozzle separators with nozzles in the rotor periphery are used for continuous outlet of the concentrate phase. This type of machine was introduced 50 years ago for yeast recirculation in distilleries. The most common application is otherwise to harvest and wash a cell mass, such as baker's or fodder yeast, with minimal losses. In the BIOSTIL process the machine also has another function. It must separate but also classify the particles in the fermentor liquid, and it plays an important role here to maintain a steady state in the fermentor.

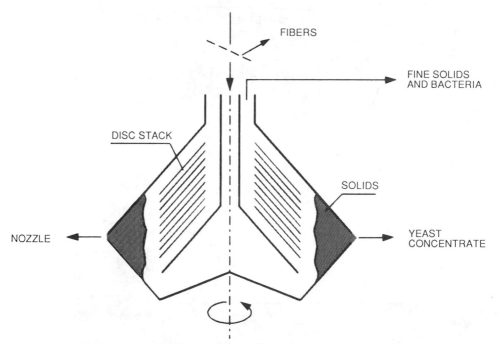

Figure 3 Schematic of high-speed centrifugal separator with peripheral nozzles.

The separation (fractionation) degree is most simply regulated by the flow or the loading of the disk stack in the rotor. Other factors that influence the separation (according to Stoke's law) are density differences, temperature, and viscosity.

The desired function must be tested on a pilot scale, however. The results can then be utilized for scaling up to industry capacities with reliable accuracy.

The machine can be used to separate yeast cells in the range of 6–8 μm or bacteria of the order of magnitude 1 μm. Furthermore, a fractionation between these can be obtained, so that yeast cells follow the heavier nozzle phase and the bacteria pass the disk stack. This technique is utilized in the BIOSTIL process to remove bacteria from the fermentor during ethanol fermentation.

Starchy raw material substrates contain large amounts of fibers and protein. The fibers are separated with different types of sieve equipment, and small fibers, protein, fat, and other small particles continue to the nozzle separator together with yeast cells. The density difference between these particles and the particle aggregates is small, but with a correct load of the disk stack the major fraction of proteins and fibers is flushed through the machine as a light phase. In other words, to recover the yeast cells a coarse separation is desirable. Much of the solids follow the yeast in the heavy phase, but a steady state is quickly obtained and sustained by drainage of the surplus. Larger centrifugal separators of this type can handle flows in the magnitude of 100 m³/h. In the fermentor a yeast content of about 6–10% by volume (2–3% DS) is maintained. By means of the separator the cells are immobilized without the risk of clogging and blocking often encountered in other systems.

C. Process Description

In practice BIOSTIL is a section in a distillation complex, and the final design of the process is dependent on the nature of the feedstock.

Feedstocks fall into three categories:

1. Sugar-based substrates
2. Starch-based substrates
3. Cellulose-based substrates

Table 1 shows the most important sections in a distillery, which are included in a complete plant dependent on raw material. BIOSTIL consists of three integrated sections: fermentation, separation, and primary distillation.

After substrate preparation two main types of substrates are obtained, often called clean substrate and whole mash.

A clean substrate means that the content of suspended particles is low; molasses from cane or beet sugar is a good example. Similarly, cane juice and starch hydrolysate contain a low content of suspended material. It is always an

Table 1 Ethanol Production: Process Sections

Raw materials	Substrate preparation				Type of Mash
	Mill	Sacch	Sugar	Conc	
Sugar					
Cane					
molasses					Clear
Beet					
molasses					Clear
Melmisto					Clear
Syrups				X	Clear
Beets	X			X	Whole
Starch					
Corn	X	X			Whole
Wheat	X	X			Whole
Barley	X	X			Whole
Cassava	X	X		X	Whole
Potatoes	X	X		X	Whole
Cellulose					
Woodchips	X	X	X	X	Clear
Straw	X	X	X	X	Clear
Paper waste	X	X	X	X	Clear
Bagasse	X	X	X	X	Clear

advantage if the fermentation can be carried out with a clean substrate, but often the costs for the substrate preparation become too high.

Whole mash means that the raw material, such as grain, is milled, saccharified, and fermented without unfermentable material being separated before the fermentor. The separation can be carried out with a higher efficiency when the sugar has been consumed and the viscosity is low. A multistage sugar extraction before the fermentation to obtain a clean substrate means either sugar losses or an unreasonable investment in extraction equipment. Therefore BIOSTIL has been developed with the capacity to handle whole mash with a high content of suspended material, usually fibers and proteins.

As mentioned earlier, the BIOSTIL technique is most efficiently utilized with a high content of fermentable material in the substrate. It is therefore economically efficient to concentrate the substrate before the fermentation. In most cases an evaporation plant is both simpler and safer, with a considerably more lenient substrate than stillage at a low pH and a high salt content.

	Biostil						
	Separation		Primary	EtOH	By-products		
Ferm	Yeast	Solids	distillation	distillation	Dry	Incineration	Other
X	X		X	X		X	
X	X		X	X		X	
X	X		X	X		X	
X	X		X	X		X	
X	X	X	X	X	X		X
X	X	X	X	X	X		X
X	X	X	X	X	X		X
X	X	X	X	X	X		X
X	X	X	X	X	X		X
X	X	X	X	X		X	
X	X	X	X	X		X	
X	X	X	X	X		X	
X	X	X	X	X		X	
X	X	X	X	X		X	

In certain cases it can be economical to mix a raw material at low concentration with one of high concentration. For instance, grain contains a high content of hydrolyzable starch. Instead of diluting with water in the substrate preparation it may be suitable to add water via a low-concentrate substrate, such as potato or cellulose hydrolysate.

Intensive work is going on with regard to cellulose-containing raw materials to find economic hydrolysis methods, acid as well as enzyme based. Generally it can be said that there is greater industrial experience with weak or strong acid hydrolysis but that the substrate is difficult to ferment, resulting in a low yield. However, the problem with enzymatic hydrolysis is the high cost of enzymes.

Raw materials, substrate preparation, fermentation, and distillation depend on the type of alcohol to be produced and its end use.

For different raw materials other streams can have varying importance from an economic viewpoint. If, for instance, sugar-containing raw materials are used, the stillage can make an economic contribution as a fertilizer. The slop or stillage from grain-based raw materials gives a high-quality animal feed product that can in certain cases give the same economic contribution as alcohol.

The BIOSTIL process is described here as applied to two typical raw materials, molasses and wheat.

1. Molasses

The design of all conventional distilleries is based on the attainment of a limiting alcohol concentration in the fermentor of around 7% wt (8–9% vol). This requires the concentration of fermentable sugars of the feed to the fermentor to be held fairly low (typically 18–27% wt), to keep within this limit.

In the BIOSTIL system the ethanol is continuously removed as it is produced, maintaining a constant level of 4.5–6% wt in the fermentor; its inhibitory effect is thereby minimized, making it possible to ferment feed streams with much higher concentrations of fermentable sugars (typically 35–60% wt).

The degree of improvement is a function of the fermentable F to the nonfermentable N ratio as shown in Fig. 2.

Table 2 illustrates the calculation of F/N for a typical cane molasses. The use of less dilution water at the front end leads to the production of a more concentrated stillage. This concentration increases with increasing F/N ratio until the osmotic pressure limitation is replaced by a stillage-handling limitation (at F/N ratios above 3.5) when the stillage concentration reaches around 60% wt (see Fig. 4).

Above this point the BIOSTIL system produces less than 1 L of stillage for every liter of ethanol produced (see Fig. 5). Because of this high proportion of organic material the stillage is autothermal; that is, it sustains combustion without additional evaporation.

Table 2 Calculation of F/N[a]

Component	% w/w	Classification
Water	25	H_2O
Sucrose (as invert)	36.0	Fermentable
Glucose	7.0	Fermentable
Fructose	9.0	Fermentable
Other organic matter	14.5	Nonfermentable
Ash	8.5	Nonfermentable
	100	

[a]F/N = 52/33 = 2.26.

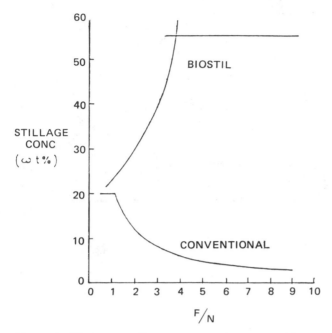

Figure 4 Maximum stillage concentration versus F/N.

With the inclusion of a suitable boiler, not only can sufficient steam be raised to run the plant but potassium-rich ash with commercial value as a fertilizer can be produced.

The process does not require preclarification or sterilization of the fermentation substrate (Fig. 6). The feed (diluted if required) is continuously metered into the single fermentor at a rate corresponding to the alcohol production rate. Nutrients are added for yeast growth.

The conditions in the fermentor are maintained at the following levels by a continuous recycling from the fermentor to the still and back.

Residual sugar	0.2% (2–3 g/L)
Ethanol	4.5–5.0% wt (6–7% v/v)
Dissolved solids	14–18% wt (18–24°Bx)
pH	Around 4.5
Temperature	Around 32°C

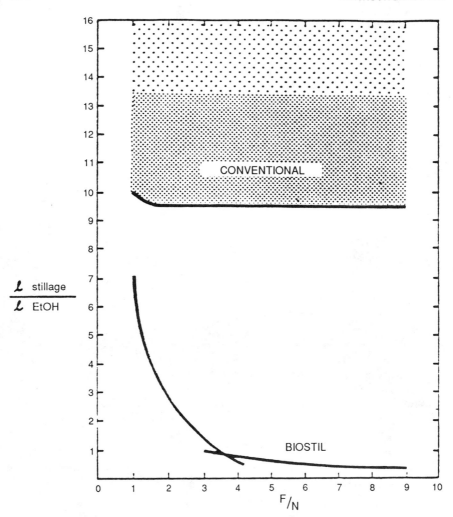

Figure 5 Liters of stillage per liter.

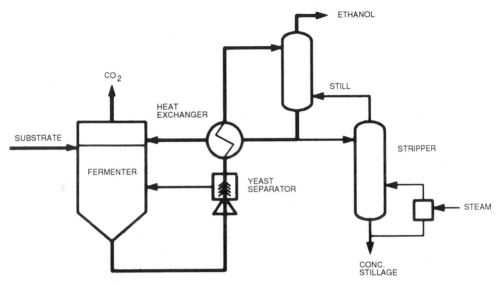

Figure 6 BIOSTIL process: ethanol production from molasses.

The heat generated during fermentation is removed by circulating the contents of the fermentor through an external heat exhanger. The high yeast concentration in the fermentor is maintained by harvesting the yeast from the recycling stream by means of a centrifuge and returning it to the fermentor. Prior to entering the still the recycling stream is preheated to around its flash point in a heat exchanger by means of the stream returning to the fermentor, which is cooled to around fermentation temperature.

The recycling stream is introduced to the top portion of the still, where the majority of the ethanol is removed at 45-50% vol. The liquid is removed partway down the still and the flow split, with the majority returning to the fermentor and the minority going to the stripping section. The split is determined by the ratio of fermentable to nonfermentable material in the feed, and it controls the quantity and concentration of stillage produced. Since only a fraction of the recirculating fluid enters the stripper column, a high vapor-liquid ratio is obtained, which enhances the column's efficiency and results in either a smaller column or reduced ethanol losses. The combination of a high vapor-liquid ratio and indirect heating of the column enables a double-effect evaporation to further concentrate the stillage.

2. Grain

As mentioned earlier, the BIOSTIL system can accept very concentrated feedstocks and this feature is exploited in the application of wheat. In fact, every ton of wheat processed requires only 0.5 ton of process water. This reduction in water permits the recovery of a dried animal feed product in a very simple process, eliminating the need for separation and evaporation. An example of an overall material balance is illustrated in Fig. 7. All four products are recovered for sale. No liquid effluent is produced.

By using wheat as a raw material it is also possible to combine the production of alcohol with starch and gluten production. The advantage of a combined production is in the lower investments and lower operating costs than when the products are produced separately. This is particularly true for wheat, since the fractionation of starch in a combined plant can be limited to the easily separated larger starch granulates.The remaining starch is then sent to fermentation. Effluent treatment is fairly limited in such a plant, since the fermentation acts as a kind of effluent treatment.

For substrate preparation, dry or wet milling systems using hammer, pin, or roller mills can be used. The feedstock flour is mixed with water and recycled material from the distillation section. A liquefaction enzyme is added, and the slurry is fed to a tank in which the temperature is maintained at 87-90°C and pH regulated at 6.0-6.5. The liquefied starch passes on to a saccharification section, where another enzyme is added to complete the conversion of starch to fermentable sugar. The saccharification is performed at 60°C and a pH of 4.5-5.0. Different conversion techniques are possible.

At the end of the hydrolysis, the "whole mash" is metered into the single fermentor at a rate that satisfies the steady-state requirements. The separation of fiber and protein then takes place within the BIOSTIL recycling loop, which

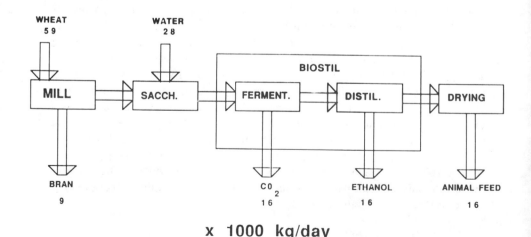

Figure 7 Overall mass balance of ethanol production from wheat.

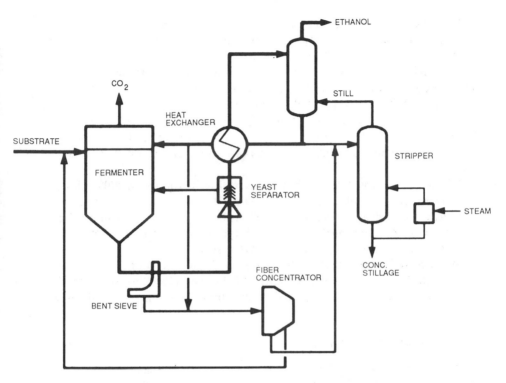

Figure 8 BIOSTIL process: ethanol production from whole mass.

gives significant advantages, such as low viscosity and no sugar losses. The practical implementation of the fiber and protein separation is illustrated in Fig. 8. Unlike the clear substrate,the stream leaving the fermentor cannot be pumped straight to the yeast centrifuge. The fiber is first separated on a bent sieve and the fiber-free phase transferred to the centrifuge, where yeast is separated and recycled to the fermentor. The fiber phase is further washed in a fiber concentration device to remove residual yeast, which is subsequently recycled to the fermentor.

The stream entering the yeast centrifuge contains both yeast and grain proteins, and it is essential that a separation be made between these two components if an excessive buildup of protein in the fermentor is to be avoided. Fortunately, yeast is slightly more dense than grain protein, and under the steady-state conditions prevailing in the BIOSTIL process, it is possible to separate more than 95% of the yeast together with only a small amount of grain protein. The bulk of the grain protein leaves the separator in the light phase and is transferred to the still column and leaves the system as a component of the stillage. This expedient ensures that the viscosity in the fermentor does not become excessive. At equilibrium the concentration of vital yeast cells in the fermentor is around 300 million per ml, and temperature and pH are maintained at 32°C and 4.5, respectively. As for other raw materials, the technique sidesteps problems of infection that plague other continuous fermentation systems.

This microbiological stability does not mean that bacteria do not grow in the fermentor, but to balance this bacterial growth the recycling of the fermentor liquid through the still column ensures a very effective destruction of bacterial cells. At equilibrium only very low levels of bacteria are present in the fermentor.

Fiber-free, deyeasted "beer" from the centrifuge is sent to the still column via the preheaters. Alcohol is stripped off as a 40–50% vol. vapor and sent on to rectification and dehydration as required. As before, a major part of the liquid is recycled back to the fermentor or to the hydrolysis section. The fiber fraction from the fiber-concentrating device goes to the stripper.

When operating on grain the concentration of the feed to the fermentor and the stillage from the stripper is controlled by solids-handling problems. The solids content of stillage being passed to the drying unit ranges around 30–35% wt.

D. Resistance to Infection

A remarkable feature of the BIOSTIL process is its resistance to infection. It does not require pasteurization of the feed, nor is acid washing of the yeast cream needed. Careful investigation has confirmed that the level of bacteria in the fermentor is kept in check by the combined effect of several process features.

First, the yeast cells are selectively separated in the centrifugal separator and are recycled to the fermentor. However, a significant portion of the bacterial cells, which are much smaller, escape separation and follow the beer stream into the still column, where they are killed by the high temperature. The recycling stream from the fermentor contains no living cells.

Second, the fermentor conditions of low pH, constantly high alcohol concentration, low sugar concentration, and high osmotic pressure seriously inhibit bacterial growth.

Third, when necessary a special yeast strain with a high osmotolerance can be used. It is most commonly used in molasses-based plants. Apart from its high osmotolerance it has good resistance to organic acids. This resistance is especially useful when substrates with high organic acid levels, such as beet molasses, are fed to the process.

E. Yield

Alcohol yields obtained in BIOSTIL projects are consistently 4–7% higher than these in conventional plants; this has been attributed to the following. The

Figure 9   Photograph of BIOSTIL plant in Brazil.

Table 3 BIOSTIL Versus Conventional Batch Plant

	BIOSTIL	Conventional batch plant
Yield, % of theoretical	94.5	87
Stillage, L/L alcohol	0.8	11
Ethanol capacity, L/day	150,000	150,000
Labor requirements	3	7
Space requirements, m^2	350	1,350

constant fermentor conditions prevailing in the process change the pattern of by-product formation relative to batch or cascade processes. A major factor is the reduced production of glycerol. In conventional distilleries, about 7–10% of the available sugar is normally converted into glycerol. In BIOSTIL, only about one-half of this quantity of sugar is lost to glycerol formation. The reduction in stillage volume leads to a corresponding reduction in the quantity of residual fermentable sugar losses. With high-purity syrups, stillage volumes and accompanying sugar losses are particularly small. Continuous process conditions with regular cleaning eliminate losses.

F. Accessibility

The process is simple to automate and has proven to be robust enough to withstand practical operation difficulties in Third World countries (Fig. 9).

IV. OTHER APPLICATIONS

The basic concept of the BIOSTIL process (see Fig. 1) can be applied to other bioconversion processes. Examples that have been tested with good results are desulfurization of coal, solvent production (i.e., acetone and butyl alcohol), continuous beer production, and continuous cell production (Table 3) (9).

REFERENCES

1. Arlie, J. P., Ballerini, D., and Nativel, F., *Rev. Inst. Fr. Petrole* 39(6, November–December) (1984).
2. Rosen, C.G., *Continuous Fermentation*. International Symposium on Ethanol from Biomass, Winnipeg, Canada, October 1982.
3. Ehnström, L., Swedish Patent No. SE 780 1133-5, Feb. 2, 1984.
4. L. Ehnström, U.S. Patent No. 44,60,687, July 17, 1984.
5. L. Garlick, *Sugar*. September: 13 (1983).

6. Cook, R., Wallnér, M., and Garlick, L., VI International Symposium on Alcohol Fuel Technology, May 21–25, 1984, Ottawa, Canada, Vol. III, pp 3–476.

7. R. Cook, Sugar y Azúcar Yearbook 1983, p. 96.

8. Alfa-Laval, Laboratory Separation LAPX 202 User's Guide, Alfa-Laval Document No. PM 40229 E2.

9. Alfa-Laval Internal Publications

10. Haraldsson, Å., and Rosén, C. G., Study of continuous ethanol fermentation of sugar cane molasses. I. System for continuous fermentation. II. Continuous alcohol fermentation and product removal in a laboratory scale plant. *Eur. J. Appl. Microbiol. Biotechnol.* 14(216, 220) (1982).

11. Rosén, C. G., Biotechnology commercial scaling-up, *Chem. Econ. Eng. Rev.* July/August(5) (1985).

12. Rosén, C. G. Biotechnology: It's time to scale-up and commercialize. *Chem. Tech.* 17(612) (1987).

Index